THE RC...

Community energy

Duncan Clark &
Malachi Chadwick

Credits & publishing information

Text: Duncan Clark & Malachi Chadwick
Design: Duncan Clark

This volume first published May 2011 by Rough Guides Ltd,
80 Strand, London WC2R 0RL

© Rough Guides Ltd

Some of the text in this book is excerpted from
The Rough Guide to Green Living by Duncan Clark,
© Duncan Clark (2009)

The publishers and author have done their best to ensure the accuracy and currency of all information in *The Rough Guide to Community Energy*; however, they can accept no responsibility for any loss or inconvenience sustained by any reader as a result of its information or advice.

No part of this book may be reproduced in any form without permission from the publisher except for the quotation of brief passages in reviews.

ISBN: 978-1-40538-652-4

Contents

Foreword 5

1. Community energy 7
Why take action on energy? 8
Why deal with energy at the community level? 10

2. Energy in the UK 15
The energy mix 15
The energy industry 18

3. Getting together 24
Starting a group 25

4. Energy saving, home by home 37
Heating & hot water 42
Lighting & appliances 51

5. Community generation: renewables 55

Renewable electricity systems 55

Renewable heat systems 66

CHP 76

6. Making it happen 77

Feasibility study 77

Fundraising 78

Design 86

Approval 88

Construction 93

Maintenance 94

7. Resources 95

Foreword

Welcome to your Rough Guide to Community Energy! At M&S, we believe sustainability makes business sense. After launching Plan A – our five-year, hundred-point eco plan – in January 2007, we saw how much we were saving by being energy efficient and using renewable energy in our stores.

We knew people across the UK were doing the same, so we launched M&S Energy, an energy tariff designed to reward households for reducing energy consumption. As well as electricity and gas, we now offer insulation and solar energy services and it's great to see how many households are taking action.

We want to go even further, though. By coming together at a community level, thousands of UK households could save money while securing energy supplies and reducing carbon emissions. This isn't just a win/win situation … it's a win/win/win!

We experienced first hand just what an impact community energy groups can have when we worked with En10ergy and Haringey Council to install solar panels on the roof of our Muswell Hill store. Now we want to make it easier for every local community energy group to get what they need. And who better than Rough Guides to combine all the information and advice in one easy-to-follow guide?

We hope you find this book useful and that it helps you and your community progress on the low-carbon journey. We'd love to hear about your progress – why not post your success stories online at http://plana.marksandspencer.com?

RICHARD GILLIES
Director of Plan A, CSR and Sustainable Business, Marks & Spencer

Pupils from Currie High School (see p.68) watch their new wind turbine being installed
Photo courtesy of Currie High School

Community energy
What and why?

Climate change, rising energy prices, economic austerity, dwindling social cohesion – the world faces a number of challenges that can often seem depressing and intractable. Scratch the surface, however, and there are plenty of seeds of hope. Across the UK and wider world, people are coming together with their neighbours and showing that, with a bit of dedication and community spirit, it's possible for ordinary people to make real progress on a whole range of big issues.

There are countless types of community projects, of course, but perhaps the most exciting at the moment are those focusing on **energy** – both saving it and generating it. The beauty of these projects is that they can help deal with so many problems at once, making a difference at the local and global levels simultaneously. This book is designed not just to explain what **community energy projects** are and to help readers set one up, but also to celebrate and showcase some of the pioneering schemes that have been launched so far. These projects should be a source of inspiration for anyone interested in getting involved in tackling climate change or building communities – and they come with a wealth of insights into what does and doesn't work.

Why take action on energy?
Climate change and energy security

Energy is in the spotlight more than ever before. That's because the way we generate and consume it has become central to a host of environmental and social issues – including **climate change**, **energy security** and **fuel poverty**. Let's look briefly at each of these.

Climate change

At present, most of our energy comes from oil, gas and coal. It's easy to understand why: these **fossil fuels** are packed full of energy and we have centuries of experience extracting and using them. Unfortunately, there is now overwhelming scientific evidence that the burning of fossil fuels is warming the planet. Largely as a result of oil, gas and coal use, humans now add more than thirty billion tonnes of **carbon dioxide** or **CO_2** to the air each year – more than four tonnes per person. All that gas goes into the Earth's atmosphere, which is surprisingly shallow: as thin as a sheet of paper wrapped around a soccer ball.

The scientific evidence points to **a rise of more than 0.7°C** over the last century. Indeed, the first six years of the twenty-first century, along with 1998, were the hottest on record. We can already see the impacts: sea ice and glaciers are dwindling, the growing season is lengthening, and many mosquitoes, birds and other creatures are being observed at higher altitudes and latitudes. (For more information on the science and symptoms of global warming, see *The Rough Guide to Climate Change*.) If something isn't done to slow the warming, civilization looks likely to face a host of potentially catastrophic problems.

Barring a technological miracle, climate change won't be solved without significantly reducing the amount of fossil fuels we consume. That, in turn, will mean reducing the amount of energy we need and generating more from renewable, low-carbon sources.

Energy security

Another good reason for changing the way we generate and consume energy is to insure ourselves against **future shortages of fossil fuels**. With a global population that is both growing and becoming more prosperous, energy demand is on the up. In parallel, reserves of the most useful and easily accessible fossil fuels are inevitably falling. In this context, governments, industry and campaigners alike are increasingly concerned about energy security. The worry isn't that the world will suddenly "run out" of energy, but rather that supplies could quickly become much more expensive and precarious – especially given that some energy-rich nations are politically unstable or unfriendly.

In particular, some experts and campaigners argue that energy prices could spiral up indefinitely thanks to **peak oil** – the point at which oil production reaches its final peak and starts to inexorably decline. No one knows for sure when peak oil will arrive or how energy markets will respond when it does, but there is a growing consensus that weaning society off fossil fuels and onto renewable sources would be sensible even were climate change not a problem.

For more information on energy security issues, see *The Rough Guide to the Energy Crisis*.

Fuel poverty

A British household is considered to experience fuel poverty if it needs to spend more than a tenth of its total income on fuel to provide adequate levels of heating. Having fallen in the years leading up to 2003, fuel poverty has subsequently increased, thanks largely to rising energy prices. Each winter, more than **four million households** – including those occupied by elderly and other vulnerable people – find it difficult or impossible to keep their homes warm. This is a major social issue, since inadequate heating doesn't just cause discomfort but can also lead to serious health complaints or even death.

Why deal with energy at the community level?
The benefits of community energy projects

Problems such as climate change and insecure energy supplies are hugely challenging. If we're to solve them, all levels of society will need to play their part, from individuals reducing their personal energy consumption to governments participating in international negotiations. Community energy projects exist between those two extremes, and it's quickly becoming clear that they have a crucial role to play. Here are just a few of the reasons why.

- **The best way to make a difference** Community-scale energy projects enable proactive citizens to make the biggest possible difference – much more than they could just by focusing on their own lifestyle or writing to their MP. In addition, community-scale projects are substantial enough to be noticed and duplicated elsewhere in Britain and wider world.

- **The right messengers** A wealth of psychological evidence suggests that successfully encouraging behaviour change and winning hearts and minds requires not just the right message, but also the right messengers. If governments or journalists suggest that people save energy, or support local wind farms, many people will switch off – or even take up the exact opposite view. Local discussion and ownership, and trusted neighbours leading by example, are likely to be far more effective in making change happen.

- **Localising energy** Community energy projects are a great way to both speed up and scale up the rollout of small- and medium-scale renewable technologies. These technologies offer a number of advantages. For example, transmission losses can be kept to a minimum, because generation takes place near the point of

COMMUNITY ENERGY

consumption. In addition, a wealth of anecdotal evidence suggests that local generation encourages people to use energy more carefully. Indeed, while efficiency measures can theoretically be susceptible to a "rebound effect" (where lower bills encourage higher consumption), there seems to be something about local generation – and the more sophisticated metering that usually accompanies it – that makes people pause for thought before flicking a switch. This amplifies the carbon savings and makes the power generated go further. It's not clear how significant this effect is or how it varies between technologies, but it bodes very well for community energy projects.

> "Before I put the washing machine on, I always look out of the window to make sure the wind turbines are spinning."
>
> A resident of Findhorn ecovillage (see p.72)

- **Efficiency** Many community projects are extremely cost efficient. Local people may contribute their time for free and local knowledge can ensure that the most suitable measures are encouraged and implemented in the right places.

Broader community benefits

The benefits of community energy initiatives go far beyond their direct impact on climate change, energy security and fuel poverty. Many communities in Britain and elsewhere are fractured by social divisions and low collective self-esteem, and community energy projects can help address this malaise. For example, a project can give normally isolated groups common ground and a **basis for trust and friendship** – not to mention a tangible reminder of their successful cooperation.

Taking control of one's own energy supply can also be highly **empowering**, boosting a sense of what's possible among individuals and the community as a whole – and bringing a host of economic benefits to boot. For example, Settle Hydro (see p.80) has not only helped the local

COMMUNITY ENERGY

MOZES in the Meadows

The MOZES project, in the Meadows area of Nottingham, is a great example of the wider community benefits of local energy projects. Separated from the rest of the city by main roads, the Meadows has struggled with poverty and crime. In 2004, a group of residents decided to see if an energy project could help restore pride to the increasingly embattled community. The result was **Ozone**, a regeneration scheme designed to help the Meadows become the first inner-city community to actively reduce its carbon footprint.

After a year, with the project foundering due to lack of proper backing from the council, some of the people involved explored the idea of establishing a community-owned energy services company to help residents cut carbon and avoid debt and fuel poverty. With startup funding and practical support from National Energy Action, MOZES (the **Meadows Ozone Energy Services Company**) was born – an inspiring initiative that gives the lie to the idea of community energy as a worthy, middle-class preoccupation.

The group was given a boost when **Nottingham Energy Partnership** stepped in with funding for a part-time energy assessor and debt advisor based at the **Meadows Partnership Trust** (MPT) – a local regeneration organisation that was to become one of MOZES' primary allies in the area. A few years on, MOZES has a number of successful projects under its belt and shows no sign of slowing down. Key to its success has been the Green Streets programme, which helped the group provide dry-lining insulation, modern boilers and energy monitors to 32 Victorian homes on the estate.

The economic profile of the area has made MOZES wary of the community share issue model preferred by many energy service companies (see p.34), but the group's zealous pursuit of grant funding paid off when they secured £650k as part of the government's **Low Carbon Communities Challenge**. This paid for **solar panels** on 55 local houses, three schools and two community buildings – in the process throwing up some technical challenges. MOZES director Julian Marsh, a local architect and lecturer, explains: "We realized we could only put PV on a few houses per street – any more and the local networks can't take it. Some people's household electrics weren't up to scratch either. For domestic PV to work the systems need to be up-to-date and properly earthed."

COMMUNITY ENERGY

CASE STUDY

Photo courtesy of MOZES

Solar PV panels on Meadows homes

from the Feed-in Tariff (see p.57) is returned to the community. Marsh sees this as an ideal use of the FIT subsidy. "The idea is to create a community endowment. The panels should bring in about £20,000 a year, which can be reinvested in future projects. We're looking forward to doing something similar with solar thermal when the Renewable Heat Incentive comes in."

The team are already seeking to broaden the scope of the project. After working with English Nature to resolve "some bat issues", they have secured planning permission for a **300kW wind turbine** on the banks of the River Trent, and are looking at their options for funding construction. The Green Streets relationship has also allowed MOZES to turn the Meadows into a test bed for future energy systems. An experimental carbon-negative house has been built, with 38 more ultra-efficient residences in the pipeline. Plans are also in place for a trial of residential **CHP** (see p.76).

Marsh encourages those looking to emulate MOZES' success to "take advantage of what's already available" in the area. In the Meadows, that included MP Alan Simpson, who recently stood down to focus on full-time environment work. "Without the help of our MP, MOZES would be nowhere."

economy, but also brought a steady flow of visitors coming to admire the project, increasing **civic confidence and pride**.

Community groups set up to work on energy issues may also help solve other local problems, too. After all, once people have worked together and demonstrated their effectiveness at changing behaviours or installing new infrastructure, they are likely to be more inclined to tackle other issues that concern them. To give just one example, the low-carbon group in Bradford-on-Avon (see p.26) attracted a host of first-time volunteers, many of whom later contributed to other projects, either energy related or not.

The political context

Not only are community energy projects inherently attractive for all the reasons discussed above. They also fit in with a broader set of cultural and political changes. For instance, localism – and in particular community empowerment and responsibility – are central to the coalition government's political programme. At the same time, reducing carbon emissions is something the UK is committed to at the national and international levels thanks to targets put in place under Labour. Put these elements together with a set of incentives designed to support micro-renewables (see chapter five) and the result is a very attractive context for community projects that save and generate energy.

Energy in the UK

How we get our energy today – and how it could change

If you're considering setting up an energy project, it makes sense to understand the wider context, including how we currently meet our needs for electricity, heat and other forms of energy. This chapter provides a quick overview. For more information, see *The Rough Guide to the Energy Crisis*.

The energy mix
Now and in the future

Like most countries, the UK currently gets the overwhelming majority of its energy from fossil fuels – **oil, coal and gas**. Only a tenth or so comes from other sources such as nuclear, wind and hydro electricity. The chart overleaf shows the breakdown of the key fuel sources, including their ratio of production to consumption in the UK. As the chart makes clear, Britain is currently a net importer of fossil fuels – which is one reason, along with carbon emissions, why locally produced renewable energy is so attractive to policymakers.

Each fossil fuel is used in lots of different ways, but oil is first and foremost consumed for transport; coal is largely used to generate

ENERGY IN THE UK

UK production and consumption of primary energy sources

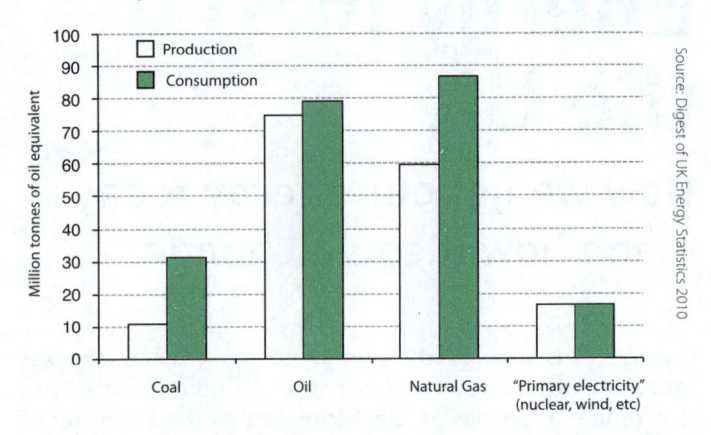

Source: Digest of UK Energy Statistics 2010

electricity; and gas is widely used for heating buildings as well as for generating electricity. As for the non-fossil sources, **nuclear** is the largest contributor overall. **Renewables** at present account for only around 3% of our total energy use and 7% of our electricity supply – and that includes sources that you might not necessarily think of as "renewables", such as burning the gas produced from landfill sites.

The future

No one can say precisely how the UK energy supply will look in ten or fifty years' time. The future fuel mix will depend on a wide range of unpredictable factors – technological, economic and environmental. However, there are already a number of legally binding targets in place that commit the UK to a particular direction of travel. These include:

- **EU energy targets** By 2020, the EU aims to generate 20% of its energy from renewable sources (as well as cutting energy consumption by 20%). As part of this effort, the UK must increase the proportion of

ENERGY IN THE UK

> ### Exploring future energy scenarios
>
> If you'd like to see how the UK's energy production and consumption could change over time, and how the various shifts could impact on carbon emissions, try experimenting with one of the online tools that let you make national-scale changes and see the results in real time. The most carefully researched such tool is DECC's 2050 Calculator, though the Guardian's National Carbon Calculator is quicker and easier to use.
>
> **DECC 2050 Calculator** 2050-calculator-tool.decc.gov.uk
> **Guardian National Carbon Calculator** tinyurl.com/guardiancarbontool

its total energy coming from renewables to 15%. To meet this target, the government is aiming to ensure that almost a third of Britain's electricity comes from renewable sources by 2020.

- **Carbon targets** The UK's groundbreaking Climate Act commits the country to an 80% cut in carbon emissions by 2050, relative to 1990. Along the way, "carbon budgets" determine how quickly we must move towards that far-off goal. The budgets agreed so far require a 34% cut by 2020 – though the government's independent body of climate change advisers have called for this interim target to be increased.

Although some campaigners claim that none of the UK's existing targets go far enough, it's clear that meeting those targets will present a huge challenge. Many experts have expressed concern, for example, that the 2020 renewables target could be missed unless progress accelerates. No one knows exactly how much community groups could contribute towards our national targets, but if they continue to grow in number and ambition at their current rate, community energy could become very significant over the coming years – especially if the groups putting in micro-renewables also help stimulate a culture of energy efficiency, as many of them are seeking to do.

The energy industry
Who controls energy supply in the UK?

The energy sector accounts for around 4% of UK GDP, making it a very substantial industry. That figure includes everything from petrol refining through to coal mining – though for the purposes of this very brief introduction to the industry, we'll focus on energy supply to buildings.

Ofgem

Ofgem is the government-run body responsible for regulation of the energy industry. As a community energy project, you probably won't have anything to do with Ofgem directly – unless perhaps you're doing something new or unusual, as in the case of Settle Hydro (see p.80). Ofgem get involved in a host of activities, including the labelling of green electricity tariffs (see p.19), but their main responsibilities are regulating energy utility companies and the monopolies that run the electricity and gas networks.

The utilities

Utilities are the visible face of the energy industry: the companies that supply each home with electricity, and in most cases gas, and bill us for what we use. In the UK, this industry is overwhelmingly dominated by a handful of large companies, usually referred to as the "**big six**": British Gas (owned by Centrica), EDF, E.ON, RWE npower, Scottish & Southern Energy (SSE) and Scottish Power. These supply their own customers and, in some cases, provide the services behind other energy offerings. For example, SSE are responsible for supplying M&S Energy customers. However, there are a number of **smaller energy providers**, too, such as Co-operative Energy, Ebico, Ecotricity, Good Energy, Green Energy UK, OVO and Spark Energy.

Some energy companies have a particular angle or target audience (for example Spark Energy is primarily for tenants) and almost all of them offer one or more electricity tariffs with some kind of green dimension.

Green tariffs

One might wonder why it's worth generating clean energy locally when it's already possible for anyone to sign up for a "**green tariff**". These, in most cases, offer to supply homes with clean power to match their consumption – which sounds ideal. However, the market for such tariffs has always been highly confusing. Friends of the Earth gave up ranking tariffs in 2004 due to the complexity of working out which were actually providing any real carbon savings, and a 2006 report by the National Consumer Council concluded that none of the available green tariffs could really claim to be offering substantial environmental benefits.

A full summary of this issue is beyond the scope of this book, but the key wrinkle relates to the fact that UK law already requires electricity providers to buy a proportion of their power from renewable sources. The cost of meeting – or trying to meet – this ever-increasing obligation is funded by a surplus on all electricity bills. This is all well and good, but it leaves open the possibility that companies can sell green tariff customers the renewable energy they were obliged to have in their portfolios anyway. If a supplier buys more than the legal minimum from renewable sources, that doesn't necessarily change things, because even a company that deals *entirely* in green electricity will still be selling the **certificates** that prove the renewable origins of the power to other energy companies, which in turn can use these to meet their own legal requirements.

Due to these kind of complications, Ofgem decided to create new guidelines for green tariffs in the form of the **Green Energy Certification Scheme**. To get certified, a tariff has to offer environmental benefits not just through the process of "providing green electricity"

ENERGY IN THE UK

OVESCo

Although the energy industry is currently dominated by straightforward *supply* companies, many commentators believe that we're starting to shift into the era of energy *services* companies, or **ESCos**. These may or may not sell energy, but their core business is the provision of services ranging from help with energy saving to the installation of micro renewables. Some of the big energy utilities see themselves moving towards the ESCo model on a corporate scale, but in the meantime a number of community-focused groups describing themselves as ESCos are shaking up energy services at the local level.

One example is OVESCo, the **Ouse Valley Energy Services Company**. OVESCo's four founders met at a Transition Lewes event in 2007 and formed an **Industrial and Provident Society** for community benefit (see p.34). Setting out to support large-scale projects, they shifted course after winning a council contract to deliver grants for household-scale renewables. OVESCo has since given out around £500,000 for more than 250 micro-renewable installations, and offered advice about home insulation. The grants have focused on **solar thermal and PV** and **wood-burning stoves** – technologies that make the most of the area's sunny climate and ample wood supplies. According to co-founder Chris Rowland, this local emphasis is central to OVESCo's success: "You've got to tailor your services to the area you're working in, which in Lewes means solar and wood fuel for domesic use. We've also built good relationships with small local companies, and I think people appreciate the peace of mind that comes with that."

With the council contract ending soon, Chris is pinning his hopes on medium-size community-owned **solar PV** installations, and at the time of writing is on the brink of a deal with a local company to lease some roof space. With growing awareness of the opportunities presented by the Feed-in Tariffs, many building owners are considering going it alone, but Chris still sees the community option as a smart commercial move. "Getting renewables installed properly involves lots of admin and a fair bit of risk", he says. "Our experience in the market means we can offer

ENERGY IN THE UK

CASE STUDY

to take care of all that much more efficiently with the backing of community investment."

Even if your group isn't hoping for a council contract, a good relationship with the local authority can be a big help. "Find someone in the council who wants to work with you", Chris advises. "Councils have to meet national emissions targets, so they're sometimes keen to help community projects raise capital." Chris would like to see the OVESCo model emulated as widely as possible, but he warns that establishing and running an ESCo isn't a casual endeavour. "OVESCo is basically a full-time job for two people, with lots of volunteers behind us […] there's a whole infrastructure to set up. Be patient, get lots of advice, and make sure you're able to put the time in."

OVESCo ovesco.co.uk
Transition Town Lewes transitiontownlewes.org

A local householder admires her OVESCo solar thermal installation

ENERGY IN THE UK

(which as discussed above raises the question of double counting) but by creating an additional environmental benefit too. This might include buying carbon offsets overseas or contributing towards community energy projects, such as solar roofs on schools.

At the time of writing, all the big six energy providers – British Gas, EDF, E.ON, npower, Scottish Power and SSE – have signed up for the guidelines for their green tariffs, along with eco specialist Good Energy. You can see their schemes, including a breakdown of the energy sources of each company, here:

Green Energy Certification Scheme greenenergyscheme.org

By contrast, Ecotricity, another specialist, has criticised the guidelines and decided to continue on the same path that it was on before: using money from its energy customers to help support its wind turbine business.

Ecotricity ecotricity.co.uk

With such a complex set of factors to consider – plus, of course, the question of the broader environmental performance of the supplier in question – it's not easy to rank the various green tariffs. Arguably the most important thing is to make sure that, if you do sign up for one, you don't succumb to the misconception that your electricity supply will be so clean that using it carefully becomes less important.

Taking a completely different approach are tariffs that focus on incentivizing **energy saving**. The premise here is that the greener you become, the less you'll pay for your power and gas. M&S Energy offer M&S vouchers to consumers cutting their annual consumption by 10%, while SSE's Better Plan offers a price cut for the same achievement.

M&S Energy mandsenergy.com
Better Plan southern-electric.co.uk/betterplan

ENERGY IN THE UK

The national grid

"National grid" refers both to the UK's electricity **transmission network** and to the company with overall responsibility for running it: **National Grid plc** (which is also in charge of Britain's gas network). The grid takes electricity from generators and transmits it to local distribution networks, where it is stepped down to mains voltage and supplied to customers. These local networks are operated by independent companies called **Distribution Network Operators** (DNOs). National Grid plc and the various DNOs are private companies strictly regulated by Ofgem.

For a grid system to function, it's not enough for the electricity to get from A to B. It also needs to be **balanced**, with electricity supply closely matched to consumption at any given moment. For that reason, grid operators have to permanently monitor the flows into and out of the system, and to be aware of every significant grid-connected generator in the country – including those owned by community projects.

Although there are plenty of off-grid energy projects, especially in remote areas, the overwhelming majority of household and community-scale generators are connected to the national grid. See p.92 for more information.

Microgrids

A microgrid takes the elements of a conventional grid and connects them at a local level. Although microgrids are often linked to the national grid, they can operate independently if necessary, protecting users from power cuts and price spikes. For obvious reasons, the microgrid model is well suited to small-scale technologies favoured by community initiatives, and although the concept is still very much in its infancy, ambitious groups may want to design their project with a future microgrid in mind.

Getting together

The benefits of local carbon and energy groups – and how to set one up

We can all cut carbon and save energy through our individual lifestyle choices – just as we can all engage with our political representatives on climate change and energy security. But if we want to make a truly tangible impact, it's usually the community level that offers the most promise. The beauty of community energy projects is their scale: small enough to be within reach for ordinary people, but large enough to make a significant difference. As the following two chapters show, there are countless forms that a project could take, but they all have one thing in common – a group of people who, in some capacity, have decided to work together for the benefit of the local community and wider world.

Community groups focusing on carbon and energy come in all shapes and sizes, from student environmental societies to well-funded coalitions of local charities, politicians, business owners and so on. The role of a community carbon reduction group in a given energy project will vary according to the capacity of the group and the scale of the

GETTING TOGETHER

project. For large-scale, capital-intensive schemes, the group usually plays a **management** role, steering proposals through the planning process, hiring contractors and coordinating the sale of shares (see Westmill Wind Cooperative case study, p.63). For less technically advanced projects, group members may play a direct role in rolling out **energy-saving measures**, as in the case of Transition Belsize and Peckham Power (see p.45). On other occasions, the need might be for **motivation or advice**. Here, the group could use its local knowledge and promotional skills to encourage residents to take advantage of existing services, and advise them on the different options available (see the Sydenham Community Insulation case study, p.48).

Most groups working on community energy projects are concerned with climate change, local regeneration or some hybrid of the two. If there aren't any established community groups in your area (or if there are, but you can't get them interested in energy), you'll need to set one up yourself.

Starting a group
The first steps

While there's no doubting the allure of an entirely new project that you can arrange exactly as you want, **setting up a group** from scratch should usually be a last resort. Whether you're going it alone or starting out with a couple of like-minded friends, it's crucial to be aware of what's already going on in your area – community time and resources are in short supply, and the last thing you want to do is use them up by reinventing the wheel. Even if you're fairly sure there are no existing groups working or willing to work on energy, thorough local research will help you make useful contacts, and may also unearth sources of advice and support.

GETTING TOGETHER

Climate Friendly Bradford-on-Avon — CASE STUDY

Four years after being set up, **Climate Friendly Bradford-on-Avon** (CFB) had built a strong local network, run a series of well-received small-scale initiatives and was ready to try its hand at something bigger. That opportunity came when the group won a **Green Streets** competition to fund energy efficiency upgrades for 98 homes, two schools and three community buildings. These renovations were designed to complement a **behaviour-change programme** incorporating everything from smart meters to education projects. One school will also get its own **solar PV** installation.

Keen to ensure the grants would reach those most in need, CFB volunteers put up posters and handed out leaflets (see picture) in poor areas and promoted the initiative in local newsletters and email lists. CFB were also determined for the project to have a "ripple effect", with grant recipients becoming local energy-efficiency advocates. Drawing on years of research into the area's housing stock and energy use, the group picked a representative range of housing types and quizzed applicants on how they would spread the word about the benefits of energy saving.

The initiative saved carbon, lowered bills and raised awareness – though it also presented challenges for the organizers. "We knew it was going to be CFB's biggest project to date", says co-founder Jane Laurie, "but we weren't prepared for the amount of admin and coordination work involved in getting the efficiency measures in place." Because the Green Streets funding was only for capital costs, the group relied heavily on **volunteers** for the delivery work. While their combined hours were equivalent to one half-time employee, "there's only so much you can ask of a volunteer", says Laurie.

CFB's long-term goal is to help the area become **carbon neutral** by 2050. Laurie knows this is ambitious, but finds reason for optimism in the community's response to their work so far. "It's important for these projects to fit into a long-term plan. For lots of the people we worked with, this was the first time they'd really thought about saving energy, but we've already seen lots of them having extra work done even after they'd spent all of their Green Streets allowance."

GETTING TOGETHER

Climate Friendly Bradford-on-Avon (see p.26) take their message to the streets

When you're recruiting people to be involved in a new project, it's important to strike a balance between having an exciting, well-thought-out idea that can inspire people to join, and leaving things sufficiently flexible and open-ended that you're able to adapt to changing circumstances and give new members an opportunity to feed in suggestions. In other words, make sure you have a strong idea, but don't get too attached to its specifics. Get this balance right, and you'll have the best chance not only of bringing people on board, but also keeping them enthusiastic and committed for the long haul.

The early days

You've dreamed up a project; now it's time to get the community involved by setting up your **first meeting**. This is your chance to grab people while they're still making up their minds, so it's worth putting in plenty of effort to ensure it goes well. Make the meeting productive, inclusive and enjoyable – and be sure that everyone leaves knowing what to do next.

GETTING TOGETHER

Your first meeting: things to think about

Beforehand

- **Promotion** As well as posters and flyers (see next page), consider using online social networks and other channels such as local newpapers and radio.

- **Venue** Your home may be suitable though risks making it seem like "your" project, so consider a more neutral location such as a local hall, cafe or pub.

- **Incentive** Pizza, wine or any other refreshments can help encourage people to come along – especially if they are free. A guest speaker, if you can find someone suitable, could also help.

- **Agenda** Although you won't want to stick to it religiously, it's always useful to have an agenda sketched out to frame the discussion.

On the day

- **Introductions** Make sure everyone knows who everyone else is. Also consider encouraging everyone to say what inspired them to come – this can be a good way to break the ice.

- **The discussion** Read out your draft agenda and let others suggest changes, then work through the items, trying to make sure everyone gets a chance to speak.

- **Wrapping up** Try to get agreement on basic principles and task a person or sub-group with writing up an initial governing document.

- **Next steps** Note down any other action points; try to set the date and time for the next meeting; and get names and contact details for everyone – this is the start of your mailing list (see p.32).

- **Follow up** Promptly circulate notes by email, confirming decisions, action points and the date of the next meeting.

GETTING TOGETHER

Promotion

Once your group is up and running, you might want to start **promoting your activities** and getting other people involved. But first have a think about who you want to communicate with, and for what purpose. Do you want people to attend events? Join the email list? Give money or volunteer time? Change their light bulbs? Or just be aware of your group's existence?

You might think you don't have the skills or experience to run a proper promotion drive, but that needn't be an obstacle. One of your members might have technical know-how, or you could seek in-kind help from a local business. For example, Climate Friendly Bradford-on-Avon (see p.26) works with a local graphic designer on a "pay if you can" basis. If that's not possible, don't give up, as many of the technical aspects of promotion have become much less demanding in recent years. User-friendly desktop publishing and fast, inexpensive printers have lowered the bar for high-quality DIY **posters and flyers**. And excellent free services such as Posterous, WordPress and Google Sites let you easily create, publish and update a **website**. Then there's the

Nicely designed posters promoting OVESCo events

GETTING TOGETHER

The Transition movement

The **Transition** movement – an international network of local groups focused on building **energy security** and tackling **climate change** – has provided fertile ground for community energy projects, including many of the ones featured in this book.

The Transition story begins in Devon in 2006, when Rob Hopkins and Naresh Giangrande founded **Transition Town Totnes**. Primarily focused on awareness-raising around climate change and peak oil (the point at which oil production peaks and start to decline), the group operated according to a set of principles that would eventually be distilled and published as the **Transition Handbook** (see p.95). Advocating a community-led response to these twin challenges, the book, combined with the founders' energetic networking, inspired groups across the UK and further afield to adopt a similar approach.

While most Transition initiatives start out with relatively small-scale projects designed to raise awareness, a handful of more established ones have put together their own 15–20 year **Energy Descent Action Plan** (EDAP) – an ambitious strategy designed by and for the local community to save, and often generate, energy.

There are several aspects of the Transition movement that make it such an effective vehicle for community energy projects. First, there's the founding concern with climate change and peak oil. A philosophical emphasis on localisation and decentralized systems also helps. But perhaps most important is the movement's commitment to finding practical, positive solutions, rather than simply campaigning against things that aren't working. That includes everything from community supported agriculture and car clubs to local currencies, neighbourhood carbon reduction clubs and urban orchards.

While many of these ideas are motivated by the need to rethink our energy systems, the Transition movement is also concerned with broader social change. The Totnes EDAP paints a picture of the town's high street in 2030 as "busy yet conversational" and describes changes to working patterns and the local economy. There's a feeling that for all

GETTING TOGETHER

Rob Hopkins at the launch of Transition Tooting in South London

its practical potential, this process of articulating a vision is partly about inspiring people and expanding the sense of what's possible. Green campaigner Alexis Rowell has this to say about his discovery of the Transition movement:

> "Suddenly I had something to say that went past the disaster-scenario planning of climate change and peak oil: I had a vision of a positive future to sell … Couldn't we imagine a better future, a greener future, a less polluted future, less materialistic, less stressed, less chemical, healthier, happier; one where there is more time to enjoy the simple pleasures in life like good food and neighbours you can trust because you know them?"

Transition groups now number in the hundreds. For more information or to find out if there's already a Transition initiative in your area, visit **transitionnetwork.org.**

constellation of **online social networks** that enable you share news, photos and videos in seconds, giving even the most casual supporter an easy way to stay informed and spread the word.

Local media is another great way to get your message out – and strong ties in this area can be hugely valuable if you find yourself caught in a planning dispute later on. Local journalists are almost universally overworked and under-resourced, so the best way to get coverage is to do the bulk of the work for them. Write **press releases** to be easily adaptable into news copy, and provide copious bright, colourful photos with prominent human faces, preferably in both portrait and landscape orientation so they can be dropped into any available gap on the page without excessive cropping.

Finding support and growing your network

Even if your group is just getting started, you're already sitting on a hugely valuable asset: your combined **contacts**. Once you've established a relatively stable core group (usually after the second or third meeting), sit down together and make a list of the individual people or networks you think might make useful allies. You may be surprised at how many key **community figures** you already have access to – MPs, councillors, headteachers, business owners, journalists, people working in community development charities, and so on. Next, make a list of links that you still need to establish, and work out how you'll make it happen. This might take some fairly ruthless networking, but it will reap dividends when you start working to get your project off the ground.

This is also the time to start developing your **email list**. Email is one of those things that's very easy to do, but much harder to do well. A few key things to consider:

- **Signing up** Give people plenty of opportunities to join your list (on your website, at all events etc) and be upfront about the frequency of messages. In deciding how much data to collect, strike a balance between making it quick to sign up and getting information (such as

GETTING TOGETHER

as where the person lives) for better segmentation later on.

- **Segmentation** This means splitting up your mailing list so people get messages that are more relevant to them. Start simple by using a separate list for your core team, to avoid bombarding casual supporters with minutiae.

- **Frequency** Nobody likes being inundated with emails, but long unexplained silences can cause people to lose interest. Unless you're in the final stages of a heated planning battle, one message a fortnight should be about right.

Photo courtesy of MOZES

MOZES director Julian Marsh (see p.12) at a local event. Email and other online channels are essential, but at the community level nothing beats face-to-face. (Fancy dress not required.)

- **Length** Less is more. It's fine to be chatty and personal, just don't take too long to get to the point. Detailed information is usually better on a blog post or webpage – which you can link to from your email.

- **Presentation** Complicated layouts can be time-consuming and often aren't compatible with every email system. Stick with simple formatting such as sub-headings to break up the text, and put the group's logo at the top to make each message recognisable at a glance.

- **Call to action** Be clear about what, if anything, you want the reader to *do*. Your call to action, if there is one, should be impossible to miss.

- **Opt-out** Hopefully most people will enjoy your messages, but it's crucial that each email contains instructions on how to opt out of receiving any more. The low-tech option is letting people know they can reply with the word "unsubscribe" in the subject line.

Funding preliminaries

Even in the early stages, before you're actually implementing any concrete plans, your group may need some funding to cover small costs such as web hosting, speaker fees or event refreshments. Your first step is to set up a **bank account** and nominate a **treasurer** to oversee it. Most high-street banks offer accounts for community groups, but before you commit to one, talk to an advisor about your plans for the group's development and make sure what they're offering will be suitably flex-

Legal structures

There are a number of legal structures for community projects and you'll probably find that different models will suit your group at different points in its development. At the outset, simply by being a group of people assembled for a common purpose, you will automatically be an **unincorporated association**. These are subject to very few legal restrictions or regulatory obligations, but ambitious groups will quickly outgrow this form as it precludes applying for significant grants, signing contracts and many other crucial elements of an energy project of any scale. For that reason, many groups move on to one of the following models – but there are many others, too, so be sure to seek professional advice before deciding what's best for your group.

- **Community Interest Companies (CICs)** Limited companies must register with Companies House and file annual returns. For their efforts, they gain the ability to hire employees, rent property, raise finance and so on – and to limit the way in which members are individually liable for any debt the group accrues. A CIC differs from a standard limited company in that it must pass a "community interest test" to ensure that its activities are not focused on private profit, and submit to what's called an "asset lock", which limits the transfer of assets and profits to other organisations. Roughly speaking, the CIC model is designed for organisations operating for community benefit but unable to register as charities.

GETTING TOGETHER

ible. It's good practice (and a basic requirement for many funders) to require two signatures for any transaction. By nominating three or four signatories to the account, you should avoid being caught out when a group member is on holiday or otherwise indisposed.

Although at first you may be able to operate on small donations and in-kind favours, access to significant funding usually requires the accountability that comes with a proper **legal structure**. There are plenty to choose from (see box, below), but the important thing is that

- **Co-operatives and community benefit societies** The **Industrial and Provident Society** has long been the legal form of choice for groups seeking community financing, but a law passed in early 2010 made the term obsolete. Shares-based societies now register as co-operatives or **community benefit societies** (CBSs). Although co-operative schemes like Westmill wind farm (see p.63) can provide community benefits, they are run in the interests of the shareholders and are free to pay interest and dividends to investors. CBSs, like Settle Hydro (see p.80), offer less in the way of private financial returns, and although a tax relief provision does provide some personal incentive, they tend to attract investors principally concerned with bringing dividends (financial or otherwise) to the wider community. A CBS can also choose to adopt an asset lock, giving public-spirited investors an assurance that their capital will not end up in private hands. Both options are also considered to be more democratic than the alternatives, with members getting one vote each regardless of how many shares they hold.

- **Charities** Charities enjoy tax exemptions and privileged access to the vast pots of money held by philanthropic trusts and foundations. Securing charitable status can be a long and demanding process, however, and successful applicants will find themselves more strictly regulated than if they'd taken the CIC route.

the group itself becomes a recognized legal entity, independent of the individuals within it.

Compared to discrete projects, **core costs** for running a group are less glamorous and have fewer tangible outcomes, making them much more difficult to fundraise for. Most core funding for community groups is managed and distributed at a local level, often through councils or foundations. For a directory of the latter, visit Community Foundations, or if you'd like detailed advice on getting your group established and making it more fundable, try Fit4Funding.

Community Foundations communityfoundations.org.uk
Fit4Funding fit4funding.org.uk

What next?

Once you have your group up and running, don't feel you should immediately be writing a planning application. Getting a major energy project off the ground can be a huge challenge, and new groups can benefit from building capacity and establishing a good local reputation before taking the plunge with a large scheme. In the words of Climate Friendly Bradford-on-Avon's Jane Laurie, "the most successful projects are slow and steady – community energy is all about the long haul". Getting to grips with the ins and outs of various types of project and the stages required to make them happen is a useful part of this preparation – which is where the rest of this book should help.

Energy saving, home by home

How to reduce the amount of energy your community needs

Although renewable energy systems are more glamorous, there are very good reasons for making efficiency and energy saving a core part of any community energy project. For one thing, reducing energy consumption is the quickest and least expensive way to reduce carbon emissions. Indeed, efficiency measures are a great way to slash bills and tackle fuel poverty. Just as importantly, the lower the energy consumption of a building or street or town, the easier it will be to generate a significant proportion of its energy locally. This chapter briefly introduces some of the ways to save energy and carbon at the household level. For more detailed information, see *The Rough Guide to Green Living*.

The carbon context

If you're advocating energy-efficiency measures in your community, it's good to have a sense of how they fit into the bigger picture. The chart overleaf shows how the UK's total CO_2 emissions break down into different activities. Roughly speaking, **home energy and personal travel**

ENERGY SAVING, HOME BY HOME

CO₂ emissions in the UK

11.6 tonnes per person

Includes international transport emissions and imported goods, but not other greenhouse gases such as methane and nitrous oxide

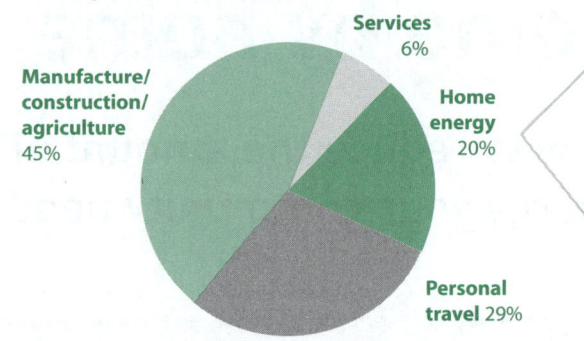

Source: Best Foot Forward

- **Manufacture/construction/agriculture** 45%
- **Services** 6%
- **Home energy** 20%
- **Personal travel** 29%

account for half of the total. Given their significance, and the fact that they are the areas over which individuals have the most direct control, these are great areas to focus on. Moreover, these are also areas where progress can be most easily measured: by keeping an eye on bills and/or distance travelled. This chapter focuses on home energy, but there's great potential for community groups to make progress in other areas, too – such as saving petrol with a local lift-sharing scheme or reducing manufacturing emissions by encouraging neighbours to share power tools and other items that take lots of energy to produce.

Different homes, different energy profiles

When promoting energy saving in the community, it's important to know roughly how much energy is consumed by different categories of homes. One factor is the **size** of a property, of course, but **age** also makes a huge difference, as the three charts on p.39 demonstrate. If your street or area largely consists of new-build properties, then it could

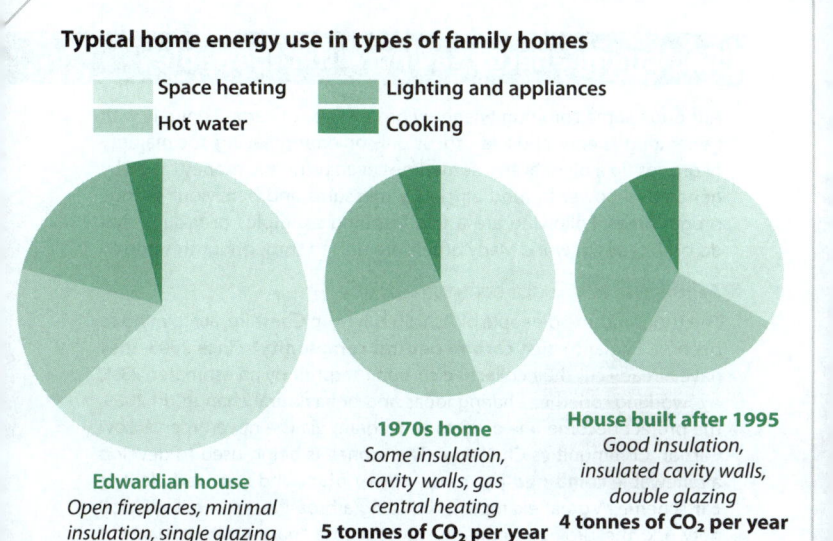

Typical home energy use in types of family homes

- Space heating
- Hot water
- Lighting and appliances
- Cooking

Edwardian house
Open fireplaces, minimal insulation, single glazing
8 tonnes of CO_2 per year

1970s home
Some insulation, cavity walls, gas central heating
5 tonnes of CO_2 per year

House built after 1995
Good insulation, insulated cavity walls, double glazing
4 tonnes of CO_2 per year

Source: Energy Saving Trust, Domestic Energy Primer

be that there is little need to improve the insulation of the buildings, and that your efforts would be better focused on behaviour change, renewable technologies and so on. If, on the other hand, you're surrounded by leaky Victorian or Edwardian properties, then insulation and draughtproofing will usually be the best areas to concentrate on.

Energy-efficiency grants

In addition to the community-scale fundraising options described in chapter six, it may also be worth investigating and promoting the ways in which individual homeowners can get financial support for energy-efficiency upgrades. For example, the government's **Warm Front**

ENERGY SAVING, HOME BY HOME

Combining energy saving with renewables

Although some community energy groups are concerned entirely with generating energy, and a few focus only on energy saving, the majority of groups do a bit of both – sometimes even using the money raised by generating power to fund efficiency measures and behaviour change programmes. Following are a few inspiring examples of groups that do both types of work. Many others are dotted throughout this book.

Ashton Hayes goingcarbonneutral.co.uk

The thousand-or-so people of Ashton Hayes, in Cheshire, are "aiming to become England's first **carbon neutral community**". Since 2006, they have already cut their collective carbon footprint by an estimated 23%, by "working together, sharing ideas and behavioural change". In 2009, the project become one of 22 to win a grant via the government's Low Carbon Communities Challenge. The money is being used to develop a renewable **combined-heat-and-power plant** and a **shared electric car** scheme. A great example of building a broad but committed local network, the project credits its success to a "hard-working band of 30–50 volunteers".

A packed event marks the launch of the Ashton Hayes low-carbon project, which is even flagged up on the sign marking the entry to the village

Photos courtesy of Ashton Hayes Going Carbon Neutral

ENERGY SAVING, HOME BY HOME

CASE STUDY

en10ergy en10ergy.co.uk

en10ergy was established in 2009 as an Industrial and Provident Society for the benefit of the Muswell Hill community in north London ("n10" is a reference to the area's postcode). Their **Low Carbon Buying Group** negotiates bulk discounts on expensive items such as **solar panels** and **efficient boilers**. And their solar project has installed panels on the roofs of local buildings – including an M&S store and a church – with the feed-in tariff revenues being used "to encourage and facilitate the reduction in carbon emissions and waste by households, businesses and community buildings in Muswell Hill and surrounding areas".

Going Carbon Neutral Stirling goingcarbonneutralstirling.org.uk

A great example of widespread community engagement, Going Carbon Neutral Stirling has worked with dozens of groups, ranging from the Stirling County Women's Rugby Team to the Stirling University Dance Society. As well as planting community herb gardens and showcasing bike-powered smoothie makers, GCNS have boosted **carbon literacy** using Mike Berners-Lee's *How Bad Are Bananas?* (see p.95) and even put 250 students at Stirling University through a "carbon detox" plan. Their Big Street Challenge is seeking to get neighbours to work together on everything from **micro renewables** to sharing of **food, tools and lifts**.

Low Carbon West Oxford lowcarbonwestoxford.org.uk

When West Oxford was hit hard by the summer 2007 floods, some local residents set up a group to cut carbon, reduce waste and traffic, and to help build "a more cohesive and resilient community." This evolved into a much-admired project that includes an **Industrial and Provident Society** as well as a **charity**. The Industrial and Provident Society, West Oxford Community Renewables, raises money from grants and share offers and uses those to fund solar, wind and hydro projects. The power is sold to the owners of the building or land, with any excess feeding into the grid. The surplus raised is donated to the charity, Low Carbon West Oxford, to pay for carbon-reducing projects relating to **energy use, waste reduction, food production** and **tree planting**.

programme provides free insulation to people on low incomes, while the forthcoming **Green Deal** scheme – due to launch in 2012 – should make it possible for all homeowners to receive energy-saving home improvements at no up-front cost. Approved installers will provide and fit the materials, and residents repay the cost over a period of many years via a surplus on their energy bills. This surplus should be more than cancelled out by the energy savings – which makes the scheme effectively free for residents.

Heating & hot water
Warming homes without warming the planet

Domestic heating accounts for more than a tenth of UK greenhouse gas emissions, and as much as half of the emissions over which individuals have direct control. There are plenty of ways to cut the carbon cost of heating, however, and some of them offer substantial financial savings, too.

Easy wins

Most households can achieve significant energy savings just by making a few adjustments to the way they use their existing heating setup. Here are some good tips to help community members get started:

- **Reducing temperatures** by just a fraction makes a big difference to energy consumption – and helps some people sleep better, too. Each degree lost saves 100–250kg CO_2 (and £20–50) per year.

- **Set boilers or tanks** to deliver water no hotter than 60°C. (Avoid going below this level, though, to ensure the system doesn't harbour the bacterium that causes legionnaires' disease.)

ENERGY SAVING, HOME BY HOME

- **Set heating programmers** carefully to match daily or weekly routines. (Homes without programmers should get one installed.) Similarly, turn the heating down or off half an hour before going out.

- **Don't heat empty spaces**. Where there are thermostats on individual radiators (see p.50), turn them down or off in rarely used rooms. If there are no radiator thermostats, consider adding them.

- **Keep your system serviced** as a well-maintained boiler is likely to be more efficient – and to have a longer life.

- **Bleed radiators** at least once a year to vent trapped air.

- **Foil reflectors behind radiators** can reflect heat back into the room. This helps rooms warm up more quickly, and – for radiators on external walls – can reduce heat loss.

- **Close the curtains** at dusk, before the warmth starts to escape. Also consider adding linings to curtains to increase their insulating power.

- **Add an insulating jacket** to any old hot water tank that doesn't already have one. Alternatively, consider upgrading to a combi boiler that heats only the water needed (see p.50).

Draughtproofing and insulation

Making our homes less thermally leaky is one of the most cost-effective ways to reduce carbon emissions and cut bills. Here are some of the options that community members might want to consider.

Draughtproofing

Draughtproofing is affordable, effective and usually manageable without professional help. There are about forty common sources of draughts and heat leakage, some of which are much easier to sort out than others. Start off with the obvious measures, such as:

ENERGY SAVING, HOME BY HOME

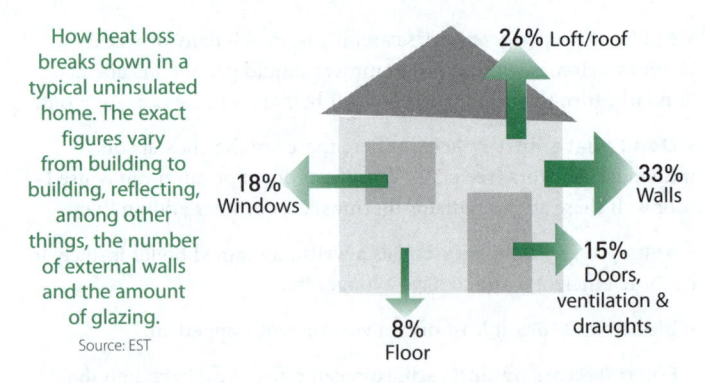

How heat loss breaks down in a typical uninsulated home. The exact figures vary from building to building, reflecting, among other things, the number of external walls and the amount of glazing.
Source: EST

- 26% Loft/roof
- 18% Windows
- 33% Walls
- 15% Doors, ventilation & draughts
- 8% Floor

- **Use chimney balloons** or specially designed panels to stop heat leaking through fireplaces. Be sure to clearly mark any blocked chimneys, however, to ensure that no-one tries to light a fire.

- **Add weatherstrips** to windows, exterior doors and loft hatches.

- **Use caulk or sealant** to fill gaps under skirting boards, around window frames and anywhere else where cold air could come in.

Draughtproofing workshops with free materials (see box, opposite) can be a great way to get the community involved in this activity.

Loft insulation

Homes with loft insulation of 10cm or less should seriously consider topping it up to at least 20cm – the current minimum requirement for newbuild properties. Doing so can save **a tonne of CO_2** per year and reduce heating bills by a fifth. Best of all, the materials are inexpensive and can usually be installed quickly and with minimal disruption – all of which makes loft insulation a great area for community groups to look at. As the Sydenham case study shows (see p.48), simply providing good information and encouraging residents to think about insulation can be very effective.

ENERGY SAVING, HOME BY HOME

Draught-busting workshops — CASE STUDY

Draughtproofing is usually manageable as a DIY project, but many people lack the confidence to go it alone – and it can be difficult to know where to begin. To help get around this inertia, Transition Belsize in North London ran a series of workshops in 2009 to help local residents become confident DIY draught-busters. After recruiting some local homeowners to host the sessions, the organizers persuaded Camden Council to provide good quality draughtproofing materials as part of a London-wide energy-efficiency pilot scheme. In this way, attendees received both hands-on practical help *and* the materials they needed to put their new knowledge into practice.

The scheme was a great example of a community approach, according to Alexis Rowell, environmental campaigner and former local councillor. "I went along to one of the workshops", Rowell says, "and was impressed by how useful and friendly it was – fifteen people passing on skills or learning new ones, and working together happily as a community."

While council funding isn't always available, other groups have found success with a different approach. South of the Thames, Peckham Power provides draughtproofing materials at cost and, for a small fee, will even come and help out with the installation.

Photo courtesy of Peckham Power

The Peckham Power team draughtproofing a local home – and a great example of a striking group logo

ENERGY SAVING, HOME BY HOME

Cavity wall insulation

Most homes built in or after the 1930s have **cavity walls** – an inner and outer wall with a gap in-between. You can usually spot cavity walls as they're relatively thick: around 30cm, compared to about 23cm for a typical solid wall. The pattern of the brickwork can also be a useful clue. As the diagram on the opposite page shows, with cavity walls, each brick is equally wide, whereas with solid walls many bricks will be placed sideways on and appear half as wide.

For homes with cavity walls, making sure the cavity is full of insulating material is usually a no-brainer. Filling an empty cavity can lead to huge energy savings for a fairly small initial outlay, and it's a quick job that causes minimal disruption. Holes are drilled in the building's exterior wall and insulation material is injected into place. If local residents aren't sure whether their cavity walls have already been insulated, encourage them to look for small circular marks left by the drilling process – or get a local insulation company to come and take a look. Note that an estimated 1.75 million homes in the UK have cavity walls that are unsuitable for filling.

Photo courtesy of OVESCo

An installer drills holes for cavity insulation as part of an OVESCo project (see p.20)

Solid wall insulation

Around a third of the UK's homes were built before 1930 and have solid rather than cavity walls. It's usually more costly and labour-intensive to insulate this kind of wall, but for some homeowners it will be the most effective step they could take to reduce their home's footprint. Solid wall

ENERGY SAVING, HOME BY HOME

Typical brick patterns of cavity walls (left) and solid walls (right)

insulation can be applied on the inside or the outside of the walls. The **internal** option involves adding boards within a wooden frame or a flexible material known as Sempatap, which rolls on almost like wallpaper. Boards give best results but also reduce room sizes. The typical cost is around £45 per square metre, plus the price of redecorating, but DIY enthusiasts can save money by doing the installing themselves – and there's no requirement to do every wall or indeed every room.

External wall insulation involves paying someone to add a thick layer of insulating render (up to 10cm deep) to the outside walls. This can be very pricy if done in isolation, though if the walls need repair anyway, the marginal cost may be as low as £1800 for a medium-sized home. Of course, exterior insulation will drastically change the look of a building – which may rule it out in conservation areas. The final finish can be flat (wet render), pebble-dashed or, for an extra cost, clad in wood, brick slip, clay or aluminium.

Double glazing

Single-glazed windows, or even old and broken double-glazing, can be a significant source of heat loss – especially in homes that have lots of glazing. New windows are usually a relative expensive way to save energy but they're well worth considering for homes that can afford it. Whatever types of windows a home-owner wants, they should look out for models with a good energy-efficiency rating. Brands can be compared at www.bfrc.org – which also offers a handy energy savings calculator.

ENERGY SAVING, HOME BY HOME

Sydenham community insulation

Built in the 1960s just outside Leamington Spa, the Sydenham housing estate is home to a diverse community living in a mix of private and council-owned homes. At its heart is the Sydenham Neighbourhood Initiatives (**SYDNI**) Centre, a local regeneration project providing services ranging from English lessons to physiotherapy. Centre manager Kate Cliffe knew that rising energy prices and leaky 60s construction made it difficult for residents to keep warm in winter, but individual attempts to improve insulation were often foiled by indifferent landlords and a tangle of conflicting information. In mid-2009, SYDNI teamed up with local charities **Action21** and **Act on Energy** to tackle the problem.

According to Rick Smith at Action21, the main issue was not that insulation was unaffordable – even for the less well-off residents – but a lack of impetus and clear information. To counter this, the newly launched **Sydenham Community Insulation Project** sought to bridge the gap between residents, landlords and insulation providers. The idea was not only to provide practical advice, but to get people talking about insulation and to nudge each other into action. Dedicated local volunteers made all the difference. "The key to the whole thing was the strength of the street champions", says Kate. "They spent a year going door-to-door and attending community events to get the word out."

Floor insulation

Some homes lose as much heat to the ground as they do through the windows. For houses with **wooden floors** on the ground level, insulating is simple: some boards are temporarily taken up and the insulation is rolled or blown between the joists. Things are trickier for homes with **solid floors**: sheets of insulating material are placed on the underlying concrete, with chipboard and a finishing layer on top. The floor will be raised, which will mean shortening doors and raising skirting boards. On the plus side, adding insulation over a concrete floor increases the speed at which rooms warm up after the heating is switched on.

ENERGY SAVING, HOME BY HOME

CASE STUDY

It was also important to engage **private landlords**, however. This group was "difficult to recruit at first", says Kate, "but once we got them on side they were able to do lots of properties at once". To counter the landlords' initial reluctance, Kate pointed out to them that tenants would be less likely to move out of warm, comfortable homes with low bills, so adding insulation could in fact reduce their administrative burden.

Despite lasting just 18 months and costing less than £10,000, the impact of the project was considerable: survey results suggest that loft insulation in Sydenham increased by 35% during the course of the programme. Thanks to the Act On Energy advice, most of these households got their insulation for free, and none paid more than £90.

Rick believes that future projects could see even greater success. "If we make the headline message really compelling, boil it down to three or four words, we'll get more people's attention and they'll look for more information themselves." He is also keen to maximise the potential of future projects to help alleviate fuel poverty. "We're looking at getting the benefits office and the job centre involved next time, and maybe contacting people who've fallen into arrears on their council tax. These are the people most in need of a warmer home and lower bills."

Boilers and heating systems

Upgrading to a **modern condensing boiler** from a ten-year-old model can slash a home's heating bills by a third – making this one of the very best ways to save energy. It works because condensing boilers extract more than 90% of the energy in the fuel, compared to around 80% for a new non-condensing boiler and just 60–70% for an older model. Community members can check the efficiency of their current boilers at sedbuk.com or quickly see whether they have a condensing model by looking outside at the flue. If it lets off steam when the boiler is in use, it's probably a condensing model; if it doesn't, it's probably not.

ENERGY SAVING, HOME BY HOME

For homes with hot water tanks, upgrading a boiler is also a good opportunity to switch to a combination ("**combi**") model. These heat water on demand rather than filling a tank with hot water that may not be used. A high-efficiency combi boiler costs around £750–1000, plus installation, but they can pay for themselves in as few as four years in larger houses. You can find a list of the most efficient boilers at:

Energy Saving Trust energysavingtrust.org.uk/Energy-saving-products

Heating controls

Modern heating controls have more **accurate thermostats**, control boilers more intelligently and offer more flexible temperature programming. Used properly, a good set of controls can cut a home's heating bills by a quarter in return for an outlay of just £150–300. Also worth considering are **thermostatic radiator valves** (TRVs). For as little as £10 each, these allow householders to control each room's temperature separately.

Lagging pipes

Where possible, it pays to insulate hot water pipes. This is best done when the pipes are installed, but even for existing set-ups it's often possible to save energy by adding inexpensive lagging to exposed hot-water pipes in lofts, cupboards and other accessible areas.

Air conditioning

A typical home air-conditioning unit uses around 1–2 kilowatts of power – about as much as a kettle or fan heater. Greener ways to keep temperatures down include **fans** and pale **reflective blinds** on south-facing windows. Good insulation also helps – as do efficient light bulbs and appliances, which kick-out less heat. A final option is to paint the roof or exterior walls white. This bounces some sunlight back into space, keeping the home cool and helping reducing global warming directly, too.

ENERGY SAVING, HOME BY HOME

Lighting & appliances
Reducing electricity use

There are many ways to reduce household electricity use – which will cut bills *and* help ensure that any locally generated power goes as far as possible. A good first step is to take a look at the chart below, which shows a breakdown of consumption in the typical home; then browse the following tips and ideas which your group may want to promote.

Electricity use in a typical UK home, excluding electric heating

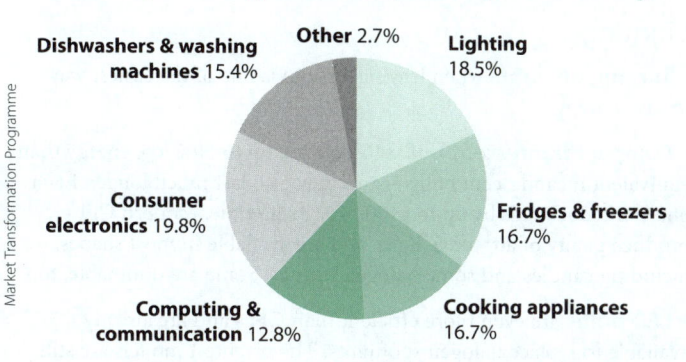

- Dishwashers & washing machines 15.4%
- Other 2.7%
- Lighting 18.5%
- Consumer electronics 19.8%
- Fridges & freezers 16.7%
- Computing & communication 12.8%
- Cooking appliances 16.7%

Source: Defra, drawing on data for the Market Transformation Programme

Kitchen

• **Fridges & freezers** run nonstop so it's well worth replacing old and leaky ones. Compared to a ten-year-old model, a new one ranked A+ or A++ for energy efficiency could pay for itself in just a few years. Consider placement, too: fridges have to work much harder if put next to cookers, boilers, hot-water pipes or sunny windows.

• **Plug-in kettles** consume than four *billion* units of electricity in Britain each year. Encouraging everyone to only boil what they need is a win-win, saving time, money and carbon.

Laundry

- **Washing machines** mainly consume energy in order to create heat, so the best way to reduce the impact of each wash – besides using a machine rated A for energy efficiency – is to select the lowest temperature that will get the job done properly.

- **Tumble drying** uses *far* more energy than washing does, so the single most effective way to save energy on laundry is to use a line or rack whenever possible. In the winter, it's best to put your rack in an empty room as the evaporation will cause a small amount of cooling.

Lighting

- **Turning off lights** when leaving a room is one of the easiest ways to save energy.

- **Compact-fluorescent bulbs (CFLs)** use up to 80% less energy than equivalent incandescent bulbs – and they also last much longer. Each one can reduce bills by up to £100 over its lifetime. Modern CFLs produce plenty of attractive light, and are available in most shapes, including candles and some halogen fittings. Some are dimmable, too.

- **LED bulbs** are even more efficient than CFLs and are already available to replace halogen spotlights. The brightest models are still relatively expensive but they save money in the long run.

Devices on standby

Although some devices in standby mode use as little as one watt (that's 3000 times less than a modern kettle) others use almost as much in standby as they do in use. Given how many devices are left on standby in the average home, this energy "leakage" is usually estimated to add up to around **5% of the electricity consumed in the typical home**. Inefficient standby modes are gradually being phased out. In the mean-

ENERGY SAVING, HOME BY HOME

Showcasing exemplar homes
CASE STUDY

An energy-saving project in a single home can take on a community aspect if the owners make the building a source of inspiration for others. A good example is Robert Cohen and Bronwen Manby's 1830s mid-terrace home in London, which recently underwent a cutting-edge eco-retrofit to add, among other things, mechanical ventilation with heat recovery; high-spec windows; solar PV; and such good insulation that the only heating systems required are an underfloor loop on the lower-ground level and a couple of heated towel rails.

An energy consultant by trade, Robert is proud of the makeover and has used open-house days to make sure others get to see it and learn from it. "I wanted to show people that it genuinely does work, but that you've got to do it well to really get the benefits", says Robert. Around fifty people visited over the course of a recent VICTERI Eco House Weekend and Robert was on hand to share tips and experiences. "The conversations were really in-depth ... people were able to relate what we'd done directly to what they might do in their own homes. It's important to understand how these things work on a human level."

To find a low-carbon "superhome" near you, enter your postcode at sustainable-energyacademy.org.uk.

Photos courtesy of Robert Cohen

A queue of visitors (right) waiting to get inspired by Robert Cohen and Bronwen Manby's eco-home (left)

ENERGY SAVING, HOME BY HOME

time, getting into the habit of turning appliances off properly – perhaps with the help of standby busters such as those made by IntelliPlug or Ivy Energy – can take a small but not insignificant chunk out of a home's electricity bills.

Ivy Energy ivyenergysaving.com
IntelliPlug oneclickpower.com

Energy monitors

Most people know roughly how much they spend on shopping but have no idea how much electricity they use. That explains why devices which display household power consumption can make such a big difference to behaviour. There are many types available, each of which has the potential to take 10–20% off a typical household's power bills. See electricity-monitor.com for a good selection.

Since energy monitors work partly by improving "**energy literacy**", using one for a few weeks can help engender longer-term change. This is handy in a community context, since individual meters can be passed around between multiple homes.

One thing a local carbon-cutting group can do is demonstrate energy-saving products, such as the ones pictured here. The ECO Kettle has a double reservoir to ensure that the correct amount of water is always boiled, while the OWL and WATTSON energy monitors make it easy for home-owners to see their electricity use in real time.

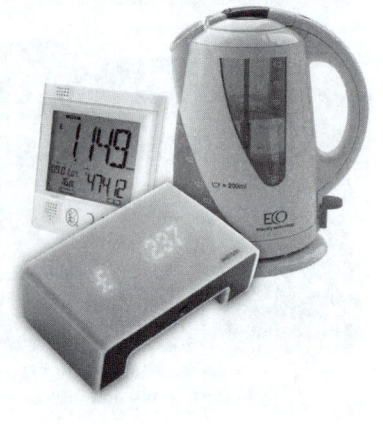

Community generation: renewables
Clean heat & power generation

With high energy prices, strong government incentive schemes and an ever-greater sense of urgency about climate change, there's never been a better time to investigate the various renewable-energy systems that can be installed at the household or community level. There are three basic categories: those which create electricity; those which create heat; and those which create both. This chapter takes a brief look at each of these categories and outlines the pros and cons of the most popular systems within them.

Renewable electricity systems
Solar PV, wind and hydro

There are many ways to generate electricity, but at the household or community scale the main options are solar photovoltaic panels, wind turbines and micro-hydro installations. Combined with a large battery set up, any of these systems can theoretically be used to enable a build-

COMMUNITY GENERATION: RENEWABLES

Understanding energy units

Energy consumption is defined in terms of watts and hours. If a 100 watt light bulb is used for ten hours, then the power consumed is 1000 watt hours. For ease we refer to 1000 watt hours as 1 **kilowatt hour (kWh)** and this is the standard unit that appears on our electricity bills. Gas bills normally specify the number of cubic metres of gas consumed but also convert the figure into kWh.

One you move up from the domestic scale, kilowatts and kilowatt hours start to give way to megawatts (MW) and megawatt hours (MWh). The "M" makes everything 1000 times bigger – so a wind turbine rated at 2MW running at full power should produce 2000 units each hour.

At the time of writing, typical prices (including standing charges) are 4p per kWh for gas and 14p per kWh for electricity. Each year, an average three-bedroom semi-detached house uses around 15,000kWh for space heating, 5000kWh for water heating and 3000kWh for lighting and appliances, costing a total of more than £1000. This results in around six tonnes of CO_2 emissions – or even more if you factor in the extraction, processing and delivery of the fuels to the home or power station.

ing or even a whole community to go "off-grid" and become genuinely self-sufficient in electricity. But in most cases that would be far more expensive and far less environmentally friendly than hooking up the generating system to the national grid. This way, when the panels or turbines are producing more power than the home or community needs – such as when everyone is out at work – the excess power is automatically fed into the grid, reducing the need for fossil-fuel generation elsewhere, and earning the person or group that owns the system some extra income. And if, for whatever reason, the system isn't generating power, the home or community can still get electricity from the grid.

Grid connections for household-scale technologies will usually be handled as a simple part of the installation package. For larger schemes, it can be more complex and expensive. See p.92 for more information.

COMMUNITY GENERATION: RENEWABLES

The Feed-in Tariff

In 2010, the UK joined many other nations in implementing a set of **Feed-in Tariffs** to give financial rewards to anyone generating electricity using small- and medium-scale technologies (anything less than five megawatts). The scheme guarantees the owners of the system a fixed amount of money for each unit of electricity generated, plus an additional 3.1p for each unit fed back into the grid.

The tariffs vary for different types of system but at any one time each rate is guaranteed and index-linked (meaning it goes up with inflation) for 20–25 years. The generosity of the scheme will gradually fall for new entrants from 2012, however, so where possible it makes sense to get projects installed as soon as possible, so that you can lock-in the most generous rates. For more on the Feed-in Tariffs, see fitariffs.co.uk.

Feed-in Tariffs for systems installed before March 31 2012
(These prices include Ofgem's April 2011 index-linked increase)

Technology	Scale	Tariff/kWh	No. of years
Anaerobic digestion	≤500kW	12.1	20
Anaerobic digestion	>500kW	9.4	20
Hydro	≤15 kW	20.9	20
Hydro	15–100kW	18.7	20
Hydro	100kW–2MW	11.5	20
Hydro	2–5MW	4.7	20
Micro-CHP	<2kW	10.5	10
Solar PV	≤4 kW (new building)	37.8	25
Solar PV	≤4kW (existing building)	43.3	25
Solar PV	4–10kW	37.8	25
Solar PV	10–100kW	32.9	25
Solar PV	100kW–5MW	30.7	25
Solar PV	Standalone	30.7	25
Wind	≤1.5kW	36.2	20
Wind	1.5–15kW	28	20
Wind	15–100kW	25.3	20
Wind	100–500kW	19.7	20
Wind	500kW–1.5MW	9.9	20
Wind	1.5–5MW	4.7	20

COMMUNITY GENERATION: RENEWABLES

Solar photovoltaics

Photovoltaic, or **PV**, solar systems contain thin layers of semiconducting material – usually silicon – that generate electricity when exposed to sunlight. The voltage from a single PV "**cell**" is low, so many cells are connected together to form a **panel or module**. These panels are in turn combined into a solar roof or some other kind of **solar array**.

Solar take-up in the UK has shot up since the roll out of the Feed-in Tariffs – especially in community projects. Without the Feed-in subsidy, PV would be offputtingly expensive per kWh generated, which has lead some commentators to argue that this technology will never make a significant contribution to UK power needs. Solar advocates counter that PV is a great way not only to generate clean energy but also to cut demand, since when solar is installed at a building, the occupants often become much more energy-aware.

Community-focused PV projects usually involve winning grants or gathering investments from community members to fund panels that are then placed on homes, community buildings or other sites, with some or all of the income from the Feed-in Tariffs being used to support other low-carbon projects. There are many such examples in this book, including MOZES (see p.12) and Low-Carbon West Oxford (see p.41). One completely different approach is a community-focused **bulk buying group** that can negotiate discounts and provide

Photo courtesy of Solarcentury

As well as panels, it's also possible to get PV tiles, designed to blend in with existing materials – ideal for some community buildings. See solarcentury.co.uk for more information.

COMMUNITY GENERATION: RENEWABLES

Howe Dell School
CASE STUDY

Howe Dell Primary School in Hertfordshire is a good example of the importance of council commitment in community-scale energy. The school originally occupied a crumbling Tudor parsonage; when it was due to move to a new development across town, the council decided that the new building should be a super-efficient eco-school full of the latest green technologies. "It was an incredibly brave decision", says headteacher Debra Massey. "They could have just put up another new school and played safe, but they believed in doing things differently."

The design and construction process took eight years, but staff and students are unanimous that it was worth the wait. The centrepiece of the project is an unusual **Interseasonal Heat Transfer** (IHT) system, with a network of underground pipes that effectively turns the playground into a giant solar panel. The asphalt warms up in the summer and the heat is transferred underground. In winter, a ground-source heat pump brings the heat back to the surface to warm classrooms, without the need for a backup gas boiler. Above ground, rooftop **solar PV** and a **20kW wind turbine** provide much of the building's term-time electricity needs, with the power exported to the grid in the holidays. And a **solar thermal** array heats water for the school kitchen.

The on-site hardware is woven into Howe Dell's "green curriculum", and it's even helping the local community grasp the new technology. "People phone up and tell me when the turbine isn't turning", says Massey. "I have to explain that it only kicks in when the wind reaches a certain speed." The school also offers a monthly "open house" event, allowing curious visitors to take a tour of what has been called the greenest building in Britain.

But while the turbine and solar panels make great teaching aids, Massey warns against creating what she calls "an overloaded Christmas tree" of impressive-looking renewable technology with no thought given to demand reduction. All of Massey's students are taught that **efficiency** comes first. At Howe Dell, high-performance windows and ultra-thick insulation keep the heat in, while natural ventilation, lightwells and photosensitive auto-dimming lights keep electricity demand to a minimum.

COMMUNITY GENERATION: RENEWABLES

support to enable more local homes to invest in their own solar system. A good example is **Energise Barnet**, a Community Interest Company in North London that focuses on PV and insulation. Find out more at energisebarnet.org.uk.

- **How much power?** A system rated two kilowatts might produce 1600 kilowatt hours of electricity per year, depending on the location (solar intensity is much greater in Dorset, say, than in Dumfries). That would be sufficient to provide the baseload of a typical home's electricity use, or about 40% of its total requirements. In most cases, the limitation for bigger systems on homes is roof space.

- **Space requirements** Each kW of panels requires around six to seven square metres of south-facing sloping roof – or a slightly larger area of flat roof or ground on which sloping frames can be erected. It is possible to install panels on other aspects and angles, though the amount of power produced will be reduced – by around 18% on a sloping roof facing east or west, or 29% on a vertical south-facing wall. Community projects sometimes make use of flat roofs on municipal, educational or commercial buildings. As well as being larger, these buildings have the advantage of having high daytime power demands.

- **Costs** The cost of solar systems varies according to the complexity and size of the installation. A typical 2.2kW domestic system might go for £10,000–12,000, including installation, though a large array on a community building would usually work out cheaper per kWh.

- **Income** At current Feed-in Tariff rates, an average household-sized system would typically earn around £750 per year through a combination of Feed-in Tariff payments and reduced electricity bills – plus the value added to the home. The system will typically be guaranteed for 25 years, though it could last much longer.

- **Planning permission** Not usually required (see p.88), though also consider the state of the electrics in the homes in question (see p.12).

COMMUNITY GENERATION: RENEWABLES

Wind power

Being blustery and endowed with long coastlines, the UK could generate a huge amount of energy from the wind. Indeed, the government sees wind power as the most important tool for reaching its renewable energy targets. So far, however, the roll out of wind power has been slower than it might have been, largely due to planning problems, of which many are the result of objections from local people (so-called "**nimbyism**"). If experience from other countries such as **Denmark** is anything to go by, community involvement and ownership could help change things, by giving local residents a stake and sense of control over proposed projects in their area. And with the Feed-in Tariffs supporting turbines of all sizes, the economics of community wind have never looked so good.

Wind turbines create electricity from the kinetic energy of moving air. The power output of each turbine depends principally on the length of the blades, the wind speed (which varies hugely by area and height) and whether there's any air-flow obstruction from buildings or trees. The bigger and higher up the turbine, the *disproportionately* better the results in terms of power generated per pound invested. Because of this, schemes focusing on **commercial-scale turbines** are arguably the most beneficial and exciting. At the other end of the scale, tiny turbines mounted on buildings will usually produce very little power – especially in urban areas with high levels of turbulence. In-between the two extremes are small- to medium-sized standalone turbines. These are less efficient than bigger models but can still produce a very useful amount of power – and they can play a symbolic role, too, especially in an educational setting (see p.69).

Community-owned wind hasn't yet taken off in the UK in quite the same way that it has in Denmark but a number of projects – such as Westmill, described overleaf – are already up and running, with many more in the pipeline.

COMMUNITY GENERATION: RENEWABLES

Westmill wind cooperative

Oxfordshire farmer Adam Twine is a patient man. At a glance, Westmill is a dazzling success, with impressive capacity, thousands of community shareholders and a string of awards. But behind these achievements is a fifteen-year saga that could easily have scuppered the project at several stages. "The whole experience has been a rollercoaster", says Twine. "It could have gone down at any point, and we're lucky to be where we are today."

Sited on a disused airfield owned by Twine, Westmill's five 1.3MW turbines are 100% community owned by 2374 shareholders based within fifty miles. The share scheme is run as a cooperative by community wind network Energy4All and advertises a return of 8%. Twine managed to raise the initial £4.5m from a single share issue in late November 2004, but an eleventh-hour U-turn from Siemens saw the turbine prices jump by 30% and the delivery date pushed back by three years – too late to comply with the terms offered to investors. Siemens eventually agreed to deliver on schedule if the co-op could raise the extra money within six weeks. Twine and his fellow directors managed to persuade shareholders to stump up another £1m and, with a loan from the Co-op Bank, the project was back on track.

Westmill's first iteration involved smaller turbines and cleared the planning committee fairly painlessly. But Twine decided to up the level of ambition when he realised that more capacity was needed to make the project economically viable. It was at this point that a small but extremely vocal "anti" campaign sprang up. Oddly for a wind farm next door to the huge Didcot power station, the opposition focused on aesthetics, though there were also concerns about vibrations harming schoolchildren and dangers posed to local gliding enthusiasts.

With thousands of shareholders, a multimillion pound budget and a major planning dispute involving both the secretary of state and the high court, Twine relied on Energy4All's experts to give him the best chance of success. "Our opponents were extremely well-resourced, so

COMMUNITY GENERATION: RENEWABLES

CASE STUDY

we needed professional help to make sure everything was technically flawless. The expert input also helped to reassure our investors."

With the planning battle won and the power flowing, the team has turned its attention to community development. A portion of the profits are diverted to the Westmill Sustainable Energy Trust (WeSET), which funds energy-related education and arts projects and runs site visits for local schoolchildren. Getting kids involved has injected a healthy dose of personality into the project: after a WeSET competition to name the turbines, the 3000 annual visitors to the site are now greeted by Wind Warrior, Gusty Gizmo, Spinner X, Huff'n'Puff and Zeus – the turbines formerly known as Westmill 1–5.

Photo courtesy of Energy4All

Turbines at Westmill, with Didcot power station visible in the background

COMMUNITY GENERATION: RENEWABLES

- **How much power?** It all depends on the size of the turbine and its location. You can check the wind speed in your area at decc.gov.uk/en/windspeed. For reference, an appropriately sited small standalone turbine (say, 6kW) with a blade diameter of 5.5m, raised 15m above the ground, should be capable of producing around 7500kWh from an average wind speed of 5mps. That's about double the needs of a typical home.

- **Space requirements** Best results are achieved at 10m or more above surrounding buildings and trees. In mounting any turbine, avoid sites with excessive turbulence, which will not just reduce performance but also shorten the device's working life.

- **Costs** Very roughly, expect to pay about £3000 per kW, including the turbine, mast and inverters. Large stand-alone turbines may also require foundation work, which can cost another few thousand.

- **Income** At current Feed-in Tariff rates (see p.57), a small to medium turbine might pay for itself in ten years or so.

- **Planning permission** The laws have become more favourable, but for bigger projects planning can still be a huge challenge. For more information, see p.89.

- **Find out more** For in-depth information on community wind, try:

Energy4All energy4all.co.uk
NEF Community Wind Guide tinyurl.com/communitywindpower

Hydro

In hydropower systems, water flowing steeply downhill – over a natural waterfall or man-made weir – is diverted via a pipe called a **penstock**. This directs the water through an enclosed turbine, which rotates to produce electricity. Systems vary in size from giant dams across whole valleys through to fairly tiny installations on steep streams running

COMMUNITY GENERATION: RENEWABLES

through people's gardens. In-between are community-scale projects such as the groundbreaking example in Settle (see p.80). Although community hydro is still fairly niche, there's a huge amount of potential. There are promising sites all over the country – many of them easy to spot by the presence of an old water mill.

- **Power, costs and income** Given the nature of hydro, every project is different, with the upfront cost and future income depending entirely on the site. So while Settle's installation cost a little over £400,000 and is expected to generate around 150,000kWh in a typical year, others may be significantly more or less expensive per unit generated.

- **Planning permission** For hydro, planning involves the Environment Agency as well as local planning authorities. See p.91.

- **More information** For much more information on hydro, explore the very good online guides at:

NEF Community Hydro Guide tinyurl.com/communityhydro
British Hydro Power Association british-hydro.org

Anaerobic digestion

Anaerobic digesters take **organic matter** – usually animal and plant waste from farms – and allow microorganisms to digest it in the absence of oxygen. The result is **biogas** (largely methane), which can be burned to create electricity or fed into the gas grid, and a nutrient-rich sludge that makes a good fertiliser. Small-scale anaerobic digestion is very popular in many countries – especially China – but it hasn't yet taken off in a big way in the UK. Many community groups in rural areas are now looking into the technology, however. To find out more about this fast developing area, visit:

Biogas Info biogas-info.co.uk
NEF Community AD Guide tinyurl.com/communityAD

COMMUNITY GENERATION: RENEWABLES

Renewable heat systems
Solar hot water, heat pumps and biomass

For most homes, heating costs more, and creates more CO2, than electricity does. So it's no surprise that renewable heating systems have huge potential to save cash and carbon. They've been attractive for some types of buildings for years – especially those which are off the gas network and reliant on expensive and polluting oil or electricity to produce heat. But low-carbon heating systems are about to receive a huge extra boost thanks to the **Renewable Heat Incentive (RHI)**, a new UK government policy that will reward anyone producing low-carbon heat, in much the same way that the Feed-in Tariffs do for electricity.

The RHI will be available for large-scale installations from autumn 2011. For each kilowatt hour of heat produced, the scheme will pay 3–4.3p for ground-source heat pumps, 8.5p for solar thermal panels and 6.5p for the combustion or injection into the gas grid of biomethane. The tariff for biomass boilers is more complex, to ensure that system owners aren't incentivized to burn more fuel than they need. It works out at 1.9–7.6p per kWh of heat, depending on the size of the system and the amount that it's used.

The domestic version of the RHI begins in 2012, though in the meantime £15 million is being made available in grants, called Renewable Heat Premium Payments, to help towards the up-front cost of installing renewable heating systems. For more information, visit:

DECC decc.gov.uk/rhi

Solar water heating

Though many people associate solar power first and foremost with generating electricity, it's also possible to convert sunlight directly into heat. Solar thermal **"collector" panels** channel the sun's energy into a building's hot water tank. Solar water heaters have been around for dec-

COMMUNITY GENERATION: RENEWABLES

ades and there are more than 100,000 installations in the UK. There are various models available, with efficiency usually increasing with price.

Solar thermal is especially attractive for buildings with high levels of hot-water demand in warmer months – such as community centres, B&Bs and campsites. But the technology can also work well on homes and will become more economically attractive with the launch of the Renewable Heat Incentive.

- **How much heat?** In the UK, a typical domestic system will provide enough hot water for a family home in the summer – and around a third of its total annual hot water demand. A home might do better that this in the sunny Southwest and less well in cloudy Scotland.

- **Space requirements** A typical three-bedroom house would need three or four square metres of roof space facing southwest, southeast or, ideally, south. The building will usually also need space for a hot-water cylinder if it doesn't have one already (though it's sometimes possible to heat the cold-water feed of a combi boiler instead). Of course, larger community buildings tend to install much larger arrays.

- **Cost and grants** Expect to pay around £3000–5000 for a household-scale system, including installation, with prices per unit of heat dropping as the system size goes up. Installation costs can sometimes be reduced if combined with other roof work.

- **Income** It remains to be seen how generous the RHI will be for domesic installations, but the income and savings for solar thermal will always vary according to summer hot-water demand and how the building previously heated its water. The savings will be highest and the payback quickest if the solar system is displacing electricity or oil rather than gas.

- **Planning permission** Not usually required. See p.91.

- **More information** A good first point of call is:

Solar Trade Association solar-trade.org.uk

COMMUNITY GENERATION: RENEWABLES

Currie Community High School

The modernist exterior of Currie Community High School in south-west Edinburgh looks unremarkable, but the exposed concrete and boxy windows disguise one of the greenest education centres in the UK. The overhaul began in 2004 with a series of energy-efficiency improvements, including draughtproofing on external doors, low-energy lighting and a new swimming-pool cover that alone saved around 100 megawatt hours of heat energy per year.

This efficiency drive coincided with work on a feasibility study for installing renewables at the school, and the formation of a steering committee – of staff, senior pupils and representatives from the Energy Saving Trust and City of Edinburgh Council – to oversee the project. After a typically protracted planning process, the committee's determination was finally rewarded in mid-2008, as work began on a **30kW solar thermal array** and an **11kW wind turbine**.

Environment Project Worker Rachel Avery attributes Currie's success to "a couple of key staff members pushing things through"; an enlightened local council looking to pilot renewables in schools; and the attraction

Photos courtesy of Currie Community High School

COMMUNITY GENERATION: RENEWABLES

CASE STUDY

of major savings (£12,500 and 69 tonnes of CO_2 a year). Two thirds of the £80,000 upfront cost was provided by the council's Sustainable Development Unit, with the other third coming from the Energy Saving Trust.

The wind turbine (pictured on p.6) has become source of pride for staff and students alike – and its output data is almost as valuable as the electricity itself. Pupils crunch the numbers as part of a green-minded curriculum that addresses sustainability in everything from art and literature to physics and maths. Work is underway for a more sophisticated real-time energy data display for the school foyer, too.

"Trying to engage young people in environmental issues is easier in primary schools", says Rachel. "It becomes less cool as a teenager to be excited about it, but lots of pupils here are involved with the Eco Committee and energy groups, and many more have taken pledges to cut their consumption at home."

Work on Currie's solar array and wind turbine (left) and the school's online energy display (below)

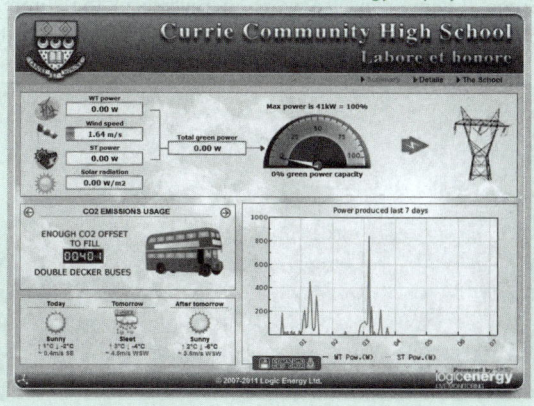

COMMUNITY GENERATION: RENEWABLES

Ground-source heat pumps

Ground-source heat pumps extract warmth from the ground. A length of plastic pipe is buried in a building's garden or land and filled with a mixture of water and antifreeze. The liquid absorbs heat from the ground (which never goes below a few degrees Celsius even when the air is freezing) and an electric compressor raises the temperature to a useful level. The heat is then distributed around the building, typically via underfloor heating, since the system is much more efficient at producing lots of low-level heat than it is at generating a smaller amount of the higher-level heat typically used by radiators.

The pump and compressor consume electricity but as long as the building is well insulated then the total energy cost and emissions will usually be far lower than with a typical heating system – especially one based on electricity, oil or solid fuel. (Of course, if you combine a ground-source heat pump with solar panels or a wind turbine, then you can create a completely renewable system.) In badly insulated buildings, the benefits are greatly reduced, because the pump has to work much harder, consuming valuable electricity.

Some systems can also work in reverse in the summer to create cooling – which is especially useful in larger buildings with high levels

The coil for a ground-source heat pump being laid in a domestic garden

COMMUNITY GENERATION: RENEWABLES

of daytime activity. Large buildings undergoing major works may also want to consider **interseasonal heat transfer**, the relatively new variation on ground-source heating installed at Howe Dell School (see p.59).

- **How much heat?** It's often possible to produce all the space heating that a building needs, and much of the hot water.

- **Space requirements** Depends on the system size. A home typically needs a trench 75–100m long and 1–2m deep. Vertical systems based on deep bore holes are possible and take up much less space, but they're also more expensive. The pump itself is a fridge-sized box.

- **Costs** A domestic-scale 8kW system costs around £6000–12,000, plus any underfloor heating. A total outlay of £10,000–13,000 would be realistic for an average house, though considerable savings can be made if major works are already being carried out on the building. Larger systems vary widely in cost.

- **Income** With the RHI (see p.66), ground-source heating is becoming a sensible long-term investment for some buildings – especially well insulated ones that rely on electric or oil heating. The rate announced for large buildings should reduce the payback time to less than a decade in the most favourable cases. It remains to be seen whether homes will received a different level of subsidy.

- **Planning permission** Not required.

- **More information** For more details on ground-source heating, see:

Ground-Source Heat Pump Club gshp.org.uk

Air-source heat pumps

Air-source heat pumps work in a similar way to ground-source models but exploit the low-level heat in the air rather than the ground. They're far less efficient than ground-source pumps for cooling (since the air gets warm on hot days), but they're almost comparable for heating and

COMMUNITY GENERATION: RENEWABLES

Findhorn's biomass boiler

With its own **750kW wind farm**, **solar hot water** and a large-scale **food growing** operation, the Findhorn Foundation's "ecovillage" status was never really in doubt. But the latest energy upgrade shows that even the most green-minded communities can always do more.

"This community is a work in progress really", says Carin Bolles, a resident who deals with the mountain of publicity generated by the project. "If you come here, you'll see different ways of doing things, and lots of experiments."

Perhaps the most ambitious of these experiments is the new **woodchip boiler**, a 250kW unit that pumps renewable heat to a handful of community buildings including Universal Hall, a large performance space and one of the village's most prolific energy consumers. Woodchips – the waste product of a nearby timber mill – are delivered fortnightly (weekly in winter) by tractor and trailer. Leftover ash is then combined with compost and used to nourish the current food crop. But an elegant fuel cycle wasn't the only reason to go with biomass: "Lots of buildings had old boilers that were burning through fossil fuels. It's a geographically compact area so district heating worked really well."

In addition to significant carbon cuts, the system saves Findhorn around £12,000 a year on fuel bills. The £300,000 cost of buying, delivering and installing the boiler was mostly covered by two large grants: £150,000 from Community Energy Scotland and £100,000 from the Energy Saving Trust. It took just three years to take the project from initial idea to grand unveiling, but it wasn't all smooth sailing. The boiler unit itself was originally supposed to be moved into position on a lorry, but a narrow access road meant the community had to bring in a 500 tonne crane (see photo) to hoist it over fifty metres of fragile duneland and mature trees.

Carin sees personal connections as the key to success in community energy: "Get to know your neighbours, first of all. Engage with people on a different level. There are enormous things that can happen when people come together and want to do things differently. It's about working with the world, rather than running away from it."

COMMUNITY GENERATION: RENEWABLES

CASE STUDY

The new system being lifted into position by a crane

Findhorn residents celebrate the new hopper by creating the logo of the 10:10 climate campaign out of wood chip

The hall, now powered by renewable heat

Photos courtesy of Findhorn

COMMUNITY GENERATION: RENEWABLES

hot water. Best of all, they're cheaper to buy (£6000–8000) and don't require the space, hassle and expense of digging a trench. The latter makes them much more suitable for flats and urban buildings. One downside is that some air-source pumps can be slightly noisy.

For a domestic system, a box around the size of a super-size suitcase is placed on the exterior of the building. The heat is either piped to a heating system or distributed directly into the rooms via fans. As with ground-source, air-source heating is best for well insulated properties.

Biomass heating and hot water

In the context of heating, biomass usually refers to **wood,** in the form of **logs, chips and pellets** (compressed sawdust). When wood is burnt, it does produce CO_2 but this is absorbed by new trees planted to replace the ones felled. That means, in effect, the carbon footprint is limited to the energy used in harvesting and transporting the wood. (Some people argue that it's actually not so green to burn wood, which should instead be used to produce furniture or buildings, locking in the carbon for long periods, but that argument is beyond the scope of this book.)

Open fires are the least efficient way to burn wood. They give out relatively little heat per log and also produce lots of soot, which contributes to the greenhouse effect. For these reasons, and to avoid having a draughty chimney when the fire isn't in use, **wood-burning stoves** are far more sensible. They are more fuel efficient, produce less soot (some can even be legally used in smoke-free zones) and some models come with an optional back boiler to provide hot water and central heating. Log stoves don't qualify for any subsidy under the RHI (see p.66), but groups like OVESCo (see p.20) have shown that stove installations can work in a community-based energy services company model.

Log-burning stoves cost around £500–1500, depending on size and quality and whether a back boiler is included. Installation is usually around £500 and, where required, a new chimney lining can add the same amount again to the total cost.

COMMUNITY GENERATION: RENEWABLES

Automated pellet and wood-chip burners

Automatic biomass heating systems are the next step up from log-burning stoves. The user fills a hopper with pellets or chips and the device takes care of itself for a few days, responding to heating controls like a regular central heating system. Pellet and wood-chip burners are yet to catch on in a huge way in the UK, but they're a proven technology. Austria alone already has more than 100,000 installations and RHI support will doubtless rapidly increase their numbers in the UK.

- **How much heat?** A larger pellet boiler system could easily produce enough heat and hot water for a standard home. Wood-chip boilers are more powerful still and best suited to large buildings with plenty of space, such as community buildings and farms.

- **Space requirements** Wood has a low energy density, so the building needs plenty of storage space. A large old house with a total water and heating demand of 25,000kWh would require roughly ten cubic meters of wood-pellet storage – or 38 cubic meters for wood chip. The building also needs space to house the boiler and hopper (generally about double the size of an oil boiler) and a flue.

- **Costs** Stand-alone room heaters cost around £1500–3000, while pellet boilers big enough for a typical house go for £5000–10,000. A wood-chip boiler for a large community building might cost £30,000–300,000, depending on the size. The project will also need to budget for fuel, of course, though pellets and chip are relatively inexpensive.

- **Income** A medium-sized community building with an annual heating requirement of 75,000kWh could potentially pay back the cost of conversion from oil to wood chip in just a few years – and the RHI will reduce payback periods yet further.

- **Planning permission** Depends on the system type.

- **Fuel** To find your nearest wood fuel supplier, visit: logpile.co.uk

COMMUNITY GENERATION: RENEWABLES

CHP
Combined heat and power

Combined heat and power (CHP) systems generate electricity and heat simultaneously. Although it's possible to run CHP on wood or some other renewable fuels, gas is currently used in the majority of installations. Using gas-fired CHP isn't carbon-free, but it's still much greener than buying electricity from the grid and creating heat separately using a boiler. That's because conventional power stations extract as little as a third of the energy from their fuel; the rest becomes heat, which is treated as a waste product and expelled via a cooling tower. Further inefficiencies occur in transmission over the grid. By contrast, CHP systems can extract almost all the energy from the fuel by creating a mixture of heat and electricity in or near to buildings that require both.

As a community energy group, there are two possible ways to approach CHP. One is to investigate the possibility of distributing or even piloting some form of micro-CHP unit. These fridge-sized devices are designed to provide heat and power to a single home. The Baxi Eco Gen is already available and many other models are in the testing phase.

Eco Gen baxi.co.uk/ecogen

The other option is to explore larger-scale systems for supplying a community building or group of homes. So-called "district heating", in which hot water is piped to homes from a local CHP plant, is increasingly popular in some countries, and it's likely that the launch of the Renewable Heat Incentive will help it catch on in the UK. Two such schemes (Ballymena and Camphill Community Glencraig) are already being planned in Northern Ireland as part of the government's Low Carbon Communities Challenge. For more on all aspects of CHP, see:

CHP Association chpa.co.uk

Making it happen

Project stages and financing

There are countless tasks involved in bringing an energy project into being – from drawing up outline plans to finding a ribbon for the local dignitary to cut at the opening ceremony. Broadly speaking, however, the process can be split into six overlapping stages: feasibility study, fundraising, design, approval, construction and maintenance. This chapter takes a brief look at each in turn, though not all of the stages will be relevant to all projects. If you're focusing on energy-saving or domestic-scale renewables, for example, you may not need to worry too much about design and maintenance.

Feasibility study
Sizing up your idea

You might know what kind of project your group is interested in, but before you can make anything happen on the ground you will need robust answers to questions such as: "is it legal?", "can we afford it?" and "how much energy will it save or produce?" This is the job of a **feasibility study**. Most groups commission one once the basic objectives for the project are agreed and there is a fairly concrete proposal (or set of alternative proposals) to decide upon. They are sober, technical documents, designed to take the project beyond the optimism and exuberance of the early planning stages and into fundable maturity.

The contents of the study will depend on the specifics of your project, but it usually makes sense to open with a background to the proposal and local area, a description of any buildings or sites in question, and details of current energy usage and supply. Next, review the proposed project, including the amount of energy that you expect to save or generate, potential financial and CO_2 savings, purchase and installation costs, income potential, lifespan, maintenance needs and regulatory issues. Finish with a summary, with recommendations and next steps. You can find inspiration by asking other groups if you can see their previous feasibility studies, or by searching online. A good example can be found at: tinyurl.com/oxfordfeasibility

A proper feasibility study takes lots of time to produce, so you may wish to start off with a quick "top-line" assessment of the project's potential – sometimes called a **screening report**.

Fundraising
Meeting the upfront cost

Like it or not, fundraising is central to any project's success, and your group's capacity for raising money will have a strong influence on what you can realistically expect to achieve. Just as importantly, success will come more easily to groups that are properly prepared for the amount of **money management** that's involved. In other words, a bit of business acumen goes a long way – which is worth bearing in mind when thinking about who you need in your core team.

Community energy infrastructure is almost always a money-saver in the long run, but the upfront costs can be eye-watering. Most projects are funded by some combination of grants, loans, shares and corporate

Offsetting and exporting

If your project will generate electricity, it's obviously worth thinking about what you will do with the power created. It could all be simply fed into the grid (this is known as **exporting**) or some of it could be donated or sold to one or more adjacent buildings (this is sometimes called **offsetting**, as the power offsets some or all of the buildings' existing electricity demands). Many community projects do both, with the exporting handled through a so-called **Power Purchasing Agreement** (PPA), a contract in which a buyer – often though not always a utility company – agrees to buy a generator's electricity for a fixed period, typically ten to twenty years. Although PPAs are a good way for companies to meet their renewables targets, you may need to be quite proactive in finding a buyer and negotiating terms. As with the other project stages, you could hire a consultant (see p.87) to take care of this, but if you do decide to go it alone, familiarise yourself with the relevant conventions, procedures and regulations to make sure you get the best possible deal.

partnerships. The best combination will vary from scheme to scheme, and it's well worth investigating all of the options even if some don't seem too appealing at first.

Grants

Grants have the considerable advantage of not having to be paid back, but this generosity tends to come with some strings attached. Grant applications can be time-consuming, with long lead times and no guarantee of a positive result. When awarded, funds often come with restrictions on when and how they can be used, and recipients must usually commit to writing a report explaining how the money was spent and detailing outcomes. Nonetheless, grants are usually crucial at one stage or another, so it's important to understand what's available and how best to access it.

Settle Hydro

Small-scale hydropower was commonplace in the Yorkshire Dales until cheap fossil fuels and the decline of local industry consigned it to history. Now, the push for renewable energy is giving this venerable technology a new lease of life in schemes such as Settle Hydro – a 50kW **reverse Archimedean screw** that sits on a weir built for watermills and even uses the original mill race to draw power from the River Ribble. A small amount of the resulting electricity is fed to the building next to the old mill, with the rest (around 90%) being sold to the grid.

The project arose from a partnership between local environmentalists (Green Settle) and a community group (Settle Area Regeneration Team). Helped along by **H_2OPE**, an organisation that supports community hydro, the coalition formed a volunteer-run **Industrial and Provident Society** (see p.34) to manage the scheme. The volunteers' commitment was soon put to the test, however, after being denied planning permission twice. The team went to the final appeal armed with a summary of every carbon reduction pledge the council had ever made, which project co-founder Ann Harding read out to the committee. That seemed to do the trick.

The £410,000 upfront cost was raised in stages. September 2008 saw the first **community share issue**. The group put a great deal of work into the share prospectus and promotion and by December they had already raised £140,000 from 166 investors. That success gave the project access to grants from environmental and regeneration bodies, which brought them up to the two-thirds level they needed for a bank loan from Charity Bank, which provided £125,000 on "massively helpful" terms. Investors, Ann says, can expect a return in the sixth or seventh year of operation, though much of the income will flow directly to community projects.

As one of the first projects of its kind, Settle Hydro's dealings with the regulatory system were frustrating. It should, for example, be receiving income from the Feed-in Tariffs, but that has been held up because Ofgem had no procedure for community hydro schemes. The Environment Agency was also unhelpful, insisting on major design changes without explanation. (To its credit, the Agency appears to have learned from these mistakes, overhauling its procedures and hiring a full-time hydro support officer.)

MAKING IT HAPPEN

CASE STUDY

By an unfortunate coincidence, the first turns of Settle's screw coincided with the start of the biggest drought since 1929, putting the first year's output at less than a third of the 165,000kWh originally projected. But the team expects to be able to make efficiency gains by lifting the screw's maximum speed (which is limited to reduce noise) and project co-founder Steve Amphlett sounds confident when he predicts 140,000kWh in year two.

Ann and Steve receive correspondence from groups all over the world seeking to reproduce their success. They consider this outreach work to be crucial but are wary of the potentially bottomless pit of work it presents. "It's our responsibility to lead by example and make sure the site always looks fantastic, but we're still just volunteers, and there's only so much we can do."

At least the project development work is out of the way, however. Juggling the legal, financial and bureaucratic elements required to get the scheme up and running took 20–30 hours of Ann and Steve's time each week, on top of already demanding full-time jobs. But the hard work was worth it, as their pride in the project is unmistakeable. "Lots of people talked about doing this but we can come back years from now and say we actually went ahead and did it … we fought the battles and cleared a lot of obstacles so others won't have to."

Ann and Steve sit next to the hydro scheme they helped create

MAKING IT HAPPEN

Community energy ticks plenty of government boxes but many of their capital grants have been scaled back recently in favour of the Feed-in Tariffs (see p.57) and Renewable Heat Incentive (see p.66): long-term output-based subsidies designed to make renewables more attractive to investors. The most notable exception to this trend is the Scottish government's **Community and Renewable Energy Scheme** (CARES), which has supported hundreds of projects in Scotland since its creation in 2009 and recently unveiled a new pot of money for farmers.

The remaining UK government support tends to come through one-off initiatives like DECC's **Low Carbon Communities Challenge**, so it's often a matter of having the right project at the right time. Keep an ear to the ground and sign up for all relevant mailing lists and discussion groups you come across. Even if they don't throw up any useful funding leads, you'll benefit from staying abreast of the latest developments in energy policy, community organising and clean technology.

The **Energy Saving Trust** maintains a database of grants for individual homes (tinyurl.com/ESThomegrants) and biggest projects (tinyurl.com/ESTprojects), but there's no need to limit yourself to green funds. A bit of lateral thinking can reap great rewards. Does your project contribute to local regeneration? Could it help out-of-work people gain new skills? Could it play an educational role? Will it help pensioners avoid fuel poverty? Get a clear idea of *all* the benefits your project offers to the local community and the wider world and you'll be able to cast your funding net as wide as possible, including the UK's huge number of philanthropic trusts and foundations. Start your search at:

GRANTnet grantnet.com
A popular search engine for grants of all types.

Lottery Funding lotteryfunding.org.uk
A portal for the funding arm of the National Lottery, which includes the Big Lottery Fund and Awards for All.

Funding Central fundingcentral.org.uk
This Cabinet Office-funded site allows you to browse or search over 4000 grants, contracts and loans. It also offers lots of good fundraising advice.

MAKING IT HAPPEN

> ### Beyond volunteering
>
> While many projects are run solely by volunteers, successfully managing a major energy project can be extremely demanding. Many of the volunteers interviewed for this book reported working up to thirty hours a week on their projects, cutting into family and social life (and indeed sleep) to keep things running smoothly. There's no doubt that it can be done, but it can make a huge difference to find funding for at least one **full-time person** to take the pressure off. There's also a lot to be said for having someone from the project available during the day to talk to contractors, regulators and other 9–5 types. An alternative approach is to try and make sure your funding covers the cost of a **consultant** who can do much of the project management on your group's behalf.

Government Funding governmentfunding.org.uk
Paid service (around £270 a year) offering a government-specific grant search and email alerts. Sometimes accused of having patchy coverage for Scotland, Wales and Northern Ireland.

European Social Fund esf.gov.uk
A large but extremely competitive EU fund focused on economic development, skills and job creation.

Loans

Financing your project through large-scale borrowing sounds daunting, but when done responsibly it can be a smart way to make up any remaining shortfall after you've exhausted other sources. As mentioned above, guaranteed long-term revenue from the Feed-in Tariffs has made renewable energy finance much easier to come by, although relying on it for loan repayments means a smaller dividend for your group and the wider community. That said, loans do offer certain advantages in their own right – in particular, lenders tend to be relatively quick to respond and less inclined to impose restrictions than grant-making bodies.

MAKING IT HAPPEN

It may be possible to get a loan from a high-street bank, but there are a few alternatives that might be a better match. The societal benefits of community energy make it an attractive field for charitable or socially conscious lenders, which offer generous lending terms for socially beneficial projects. There are many such lenders out there, the best known of which include the Charity Bank, Triodos and the Co-operative. For more, including region-specific institutions, try Finding Finance.

Charity Bank charitybank.org
Triodos Bank triodos.co.uk
Co-operative Bank co-operativebank.co.uk
Finding Finance findingfinance.org.uk

Issuing shares

The community shareholder model is a relatively recent and extremely promising development in renewable energy financing. Local people fund a project by buying shares, in return for which they (or the community as a whole) gets a portion of the revenue and a say in how the whole thing is run. See p.34 for an overview of the different legal structures available for community share ownership.

Share offers have proved remarkably effective, contributing a substantial proportion of the capital for many of the UK's largest and most successful community energy projects. As a bonus, when the community can literally buy into what you're doing, local acceptance gets a boost – which can help with planning and other issues.

There's no simple recipe for a successful share offer, but a clear and compelling proposition, tireless, savvy promotion and diligent relationship management with prospective investors will go a long way. If you're thinking about going down this route, Community Shares (communityshares.org.uk) provides an indispensable set of resources, including in-depth research on the profile, motivations and experiences of community shareholders, a briefing on legal structures and governance, and a guide to producing share offer documents.

MAKING IT HAPPEN

Business partnerships

Some large companies – particularly energy companies – have recently started to support community energy projects in one way or another. While many schemes operate on a grant-giving basis, they differ from philanthropic funding in their emphasis on publicity: companies are usually keen to maximise the PR benefits of their involvement, so expect to pose for a picture or two.

Among the larger energy companies, **E.ON**'s Sustainable Energy Fund offer capital grants, while **npower** recently repurposed the fund linked to its Juice green tariff to help schools and other community buildings get biomass boilers. **British Gas**, meanwhile, has supported efficiency and renewable projects to the tune of £2m via its **Green Streets** programme. Green Streets is now coming to an end, however, with the company's emphasis apparently shifting to a new initiative called EnergyShare (see energyshare.com).

As for the smaller players, **Good Energy** has community energy at the core of its business, as the electricity it offers to its customers is sourced through Power Purchasing Agreements (see p.79) with independent generators. As well as buying and selling community-generated power, Good Energy sometimes provides financial support to projects in the development stage. It is, for example, a major shareholder in the Bro Dyfi wind project in Powys, Wales.

But when considering a corporate partnership, don't limit yourself to energy companies. As the profile of community energy rises, it's likely that a wide range of businesses and social enterprises will become more interested in supporting local energy projects. There's even a mobile phone operator, **CMobile,** that donates 20% of its profits to community energy schemes. For more information, see: cmobile.co.uk

Pro bono help

While not providing direct capital finance per se, there's a growing coalition of businesses offering services to community energy projects on a pro bono (free) basis. Many are signed up with Carbon Leapfrog, a charity that connects them with projects in need. Well-targeted pro bono assistance can substantially ease the fundraising burden by reducing or removing the cost of otherwise expensive services such as accounting, law, engineering, PR, banking and consultancy.

Carbon Leapfrog carbonleapfrog.org

Design
Drawing up plans

For projects that involve new infrastructure, design is a key stage – or, more commonly, two stages. A basic **outline design** may be drawn up as part of a feasibility study or – more commonly – after funding is in place. This design will be crucial for getting permission from the local planning authorities and any other relevant regulatory bodies. Once permission is granted, a **detailed design** will then need to be drawn up, incorporating any changes required by the planning authorities and including sufficient amount of detail for contractors to work from.

What's required at each stage depends on the technology and site in question. But a fairly typical outline design will involve the following:

- A plan of the **current site**
- A plan of the proposed **finished site**
- Designs of any **new structures**
- Designs for **new electrical systems**
- The **specifications** of any major hardware components

By contrast, a detailed design will include precise measurements and specify materials and other details required for construction. This can be an expensive and time-consuming process, which is one reason why it's not usually initiated until the project has a green light to proceed.

MAKING IT HAPPEN

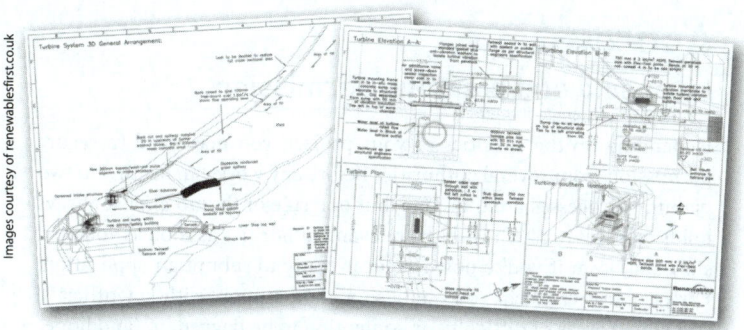

Extracts from an outline design (left) and a detailed design (right) created by hydro and wind consultancy RenewablesFirst

Consultants

At any stage of your project – but especially the design and planning phase – you may need some expert help. As discussed, it's sometimes possible to get certain kinds of advice and services pro bono, and there are **support groups** for some specific types of project. For example, Energy4All (energy4all.co.uk) provides services to wind-energy cooperatives and is owned by the groups that it supports. But there are also a large number of **private consultants** offering help to community energy groups with everything from design and planning to installation and maintenance. These range from small companies specialising in one or more technologies to the renewable departments of larger firms.

Good consultants can be expensive but they bring technical expertise, industry knowledge and – ideally – experience of similar projects. Unless those skills already exist within your team, a consultancy firm can save a great deal of time and give the project a much better chance of success. In some cases it will be near impossible to get a project off the ground without one.

It's easy to find a range of consultants via Google, but where possible try to get advice from other groups that have run similar projects.

MAKING IT HAPPEN

Approval
Licensing and planning permission

Depending on the specifics of your project, you may need to secure permission from several authorities before any work can begin. No two planning processes are the same, but one rule of thumb almost always holds true: get in touch with the relevant authorities as soon as you possibly can, even if you're nowhere near ready to submit an application. Early contact can help you establish a good rapport with key officials, and allows areas of difficulty or contention to be flagged up (and hopefully resolved) before the formal process begins.

Planning permission

Planning law is designed to ensure that new developments don't have an unduly negative impact on the amenity of the surrounding area. With its strange mix of detailed scientific analysis and volatile local politics, it can be unpredictable territory for community energy projects.

Planning permission in the UK is governed by local planning authorities (LPAs), which are usually divisions of the district or borough council. In considering an application, the LPA bases its decision on a consultation with local stakeholders and organisations with particular interest or expertise relating to the proposal. These might include bodies like English Heritage, the Environment Agency, the MOD and Natural England.

You can find an overview of the UK planning system, as well as up-to-date information on the planning and building regulations governing energy-related developments, at the UK Planning Portal, but following are the key points for each main type of project.

UK Planning Portal planningportal.gov.uk

MAKING IT HAPPEN

> ## Environmental Impact Assessments
>
> The purpose of an **Environmental Impact Assessment** (EIA) is, as the name suggests, to anticipate the potential environmental impacts of a proposed project and to identify ways in which these impacts can be mitigated. Most community energy projects are small enough to fall outside EIA legislation altogether and those that don't are usually classed as **Schedule 2** projects, meaning they only require an EIA if they are individually "judged likely to give rise to significant environmental effects". (Schedule 1 projects are things like airport runways and nuclear power stations.) Projects listed under Schedule 2 include geothermal drilling, hydro generating more than 500kW and wind installations incorporating more than two turbines or any turbine more than 15m tall.

Wind

As tall structures with large moving parts, it's no surprise that wind turbines can have a rough ride through the planning process. Following are the main issues you'll need to consider. Bear in mind that government's Planning Policy Statement for Renewable Energy (PPS22) states that "Local Planning Authorities should satisfy themselves that such issues have been addressed before considering planning applications."

- **Birds** Your application will be expected to consider the risk of birds colliding with the turbines in addition to other types of possible disturbance and habitat loss. However, the RSPB has been supportive of action on climate change and has a constructive stance on wind energy. Frequently consulted during the planning process, it opposes just 7% of UK applications on the grounds of their impact on birds.

- **Bats** In monitoring the impact of wind turbines on bats, Natural England plays a comparable role to that of the RSPB for birds. They recommend a buffer zone of around 50m between the closest turbine and any "habitat features" such as trees and hedges. Researchers

MAKING IT HAPPEN

are investigating ways to further reduce potential impacts, such as deactivating turbines during bat migration periods.

- **Military operations** The Ministry of Defence (MOD) has a dedicated wind energy team representing its interests in the planning process. That's because large turbines can potentially cause complications with radar, seismology, communications and low flying. The MOD asks to be consulted about any proposed turbine standing more than 11m tall or with a rotor diameter of more than 2m. As always, it's good to make contact sooner rather than later.

- **Civil aviation** As a user of radar across the country, the civil aviation industry – including airports, air-traffic control and en-route services – has a stake in wind farm development. The industry's umbrella organisation, the Civil Aviation Authority, recently withdrew from its voluntary involvement in the planning process for wind farms. Because of this, your best bet is to ask your local planning authority for advice on which aviation bodies you need to consult.

- **Noise** Noise issues can usually be mitigated through intelligent design and siting. A good feasibility study will consider noise disruption to nearby buildings, covering proximity, window orientation and even the noise attenuation potential of the structure's outside walls. An important (if widely criticised) document is the 1997 ETSU report on turbine noise, which the government treats as the definitive statement on the issue.

- **Visual impact** Visual objections can easily derail a wind project. If a project sited on nondescript farmland in the shadow of a giant coal-fired power station can be opposed on aesthetic grounds (see p.62), it's fair to say there are no safe bets. Local development policies usually contain area-specific guidelines on the minimization of visual impacts through turbine siting, layout and design.

MAKING IT HAPPEN

Solar

Solar PV and thermal panels are classed as "permitted developments" (i.e. they don't require planning consent) on sloping residential roofs. The exceptions are homes in conservation areas wanting to install panels visible from the road, and any kind of listed building. Solar installations on non-residential properties or positioned on flat roofs or the ground *do* need planning approval. Get in touch with your LPA to check your status and initiate an application if necessary.

Hydro

In addition to the usual planning process, most hydro projects need abstraction and/or impoundment licenses from the Environment Agency or Scottish Environmental Protection Agency. These basically give you the right to take water from the river and include conditions regarding flow depletion, protection of fish and the wider ecosystem, and the preservation of other river amenities, such as canoeing access. Hydro projects also require Flood Defence Consent, which regulates the flood risk posed by the construction and ongoing operation of the scheme. Environmental licensing is usually the first stage of the approval process, as EA/SEPA consent is essentially a precondition for any successful hydro planning application.

Biomass

Biomass boilers are classed as permitted developments if the entire installation will be housed inside an existing building, and if any new flue isn't unduly conspicuous. Commercial-scale boilers are classed as "notifiable work", meaning the local building authorities must be informed before work begins. You'll also need to consider the traffic impact of regular fuel deliveries and make sure the boiler model and fuel type are in line with local air-quality regulations.

Heat pumps

Ground- and water-source heat pumps are classed as permitted developments – with the usual caveats for listed buildings and conservation areas. Air-source heat pumps are currently in legislative limbo: they may soon receive permitted development status, but in the meantime you'll need to contact your LPA to find out the latest.

Insulation and efficiency

Energy-efficiency measures are usually only of interest to the planning authorities if the building in question is individually listed or within a conservation area or World Heritage Site. Even then, the rules only apply to changes that would "unacceptably alter the building's character or appearance". These restrictions can limit the options for double glazing, external wall insulation, boiler flues and major structural work, but even within a conservation area it's usually possible to substantially improve the efficiency of a period building (see p.53).

There are also standard building regulations to bear in mind, but with good contractors this should pretty-much take care of itself.

Grid connection

Electricity grid connections in the UK are the responsibility of the local **Distribution Network Operator** or **DNO**. The UK is split into fourteen zones, each of which has a monopoly DNO that is run as a private company regulated by Ofgem (see pages 18 and 23). As with the planning process, it's sensible to make contact with your DNO long before you're ready to formally request connection, as they'll be able to advise you on any area-specific issues.

The cost of connecting a project to the grid is determined by two main factors: the distance between the project and the nearest connection point, and whether any major system upgrades are necessary for a safe connection. If the project's output exceeds local grid capacity (which happens quite a lot in Scotland, especially), expect extra cost and a delay.

Your local DNO can provide a free connection quote, which will distinguish between "non-contestable" work that must be carried out by the DNO itself and "contestable" work that can be assigned to another contractor. If you're happy with the quote and are ready to proceed, the DNO will take a deposit (usually around 25% of the cost of any non-contestable works) and place you in a queue to get the work carried out.

Construction
From idea to reality

Contractors are the ones who pour the cement, drill the holes and lay the cables that bring your project into being, so it's crucial that they're recruited, briefed and managed effectively. One approach to finding the right people is to ask and call around – ideally speaking to groups that have commissioned similar projects. Another is to use a specialist contractor recruitment agency, or to favour an energy consultancy firm (see p.87) that will take care of this element as part of a wider package. If you do opt for an agency, try to find one that has experience working on community energy projects that involve your chosen technology.

However you plan to find your contractors, try to get a sense of the current market rate for the services you need before you start. This is usually best done through colleagues or contacts who have experience in the area, as the rates quoted online or on adverts can be misleading. Once you've found your contractors, it's helpful to remember that they are not employees. The best ones are in high demand so it's important to cultivate a good relationship, particularly if you're pleased with their work and think your group might run similar projects in future. Brief them thoroughly, stay on top of invoices and resist the temptation to micromanage, and this shouldn't be a problem.

The construction process often throws up issues that need a quick decision, so if you have a large group it might make sense to delegate responsibility for day-to-day **construction management** to a small task force of the most qualified and interested people. There's no substitute for first-hand experience, so if you don't use an agency or consultant, it may be useful to try and find someone to help out on a voluntary basis, either through your personal contacts or the group's mailing list (see p.32). Even if you only find someone to provide a half-hour run-through of the key dos and don'ts, it might be well worth the effort.

If you need to brush up on the basics of hiring contractors, you'll find a useful introductory guide at:

Contractor Calculator contractorcalculator.co.uk

Maintenance
Keeping it running

After putting in all that work, your group will want to make sure that any hardware you have installed stays in good condition. Upkeep requirements differ drastically between technologies. Most solar panels need nothing more than the occasional once-over with a brush to keep them clear of leaves and other debris, whereas biomass boilers need to have their ash collectors emptied and plates cleaned on a regular basis. Systems of all types usually come with a manufacturer's warranty, to cover repairs for between one and 25 years (some systems have different warranties for individual parts). Outside the warranty period, servicing and repairs are usually arranged through a maintenance contract. Until fairly recently, manufacturers effectively had a monopoly on these, but a maturing market has prompted some third-party firms to start offering their own maintenance schemes.

Resources
Where to find out more

This book is dotted with suggestions for websites where you can find out more on the topics being discussed. Here are some further sources that might help you find inspiration and support for your project.

Books

Communities, Councils & a Low-Carbon Future by Alexis Rowell

How Bad are Bananas? The Carbon Footprint of Everything by Mike Berners-Lee

Local Sustainable Homes by Chris Bird

Ten Technologies to Save the World by Chris Goodall

Sustainable Energy Without the Hot Air by David MacKay

The Rough Guide to Climate Change by Robert Henson

The Rough Guide to the Energy Crisis by David Buchan

The Rough Guide to Green Living by Duncan Clark

Websites

RESOURCES & INSPIRATION

Community Central
communitycentral.co.uk

Community Energy Online
ceo.decc.gov.uk

Community Energy Scotland
communityenergyscotland.org.uk

Energyshare
energyshare.com

EST Green Communities
energysavingtrust.org.uk/cafe

Greening Campaign
greening-campaign.co.uk

Low Carbon Communities Network
lowcarboncommunities.net

RESOURCES

NEF community energy guides
nef.org.uk/communities

PlanLoCaL planlocal.org.uk

GROUPS & ASSOCIATIONS

Communities & Climate Action Alliance tinyurl.com/CCAalliance

Community Energy Practioners' Forum tinyurl.com/CEPForum

ENERGY SAVING & CUTTING CARBON

10:10 1010uk.org

Energy Saving Trust
energysavingtrust.org.uk

RUNNING A GROUP

Mid Sussex Council's guide to setting up a community group
tinyurl.com/groupguide

DECC guide to community energy legal structures
tinyurl.com/DECCguide

LEGAL ADVICE

Carbon Leapfrog carbonleapfrog.org

FUNDRAISING

Fit4Funding
fit4funding.org.uk

Climate Challenge Fund Scotland
keepscotlandbeautiful.org/ccf.asp

Co-operatives guide to community investment
tinyurl.com/coopinvestmentguide

Community Shares
communityshares.org.uk

PLANNING & LICENSING

Planning Portal
planningportal.gov.uk

Government Planning Policy Statement on renewable energy
tinyurl.com/planningps22

Environment Agency Hydropower
environment-agency.gov.uk/hydropower

Natural England technical note on bats and wind turbines
snh.gov.uk/docs/C245244.pdf

English Heritage planning advice
tinyurl.com/ehplanning

MOD wind turbine statement
tinyurl.com/MODwind

SEPA planning guidelines
sepa.org.uk/planning/energy.aspx

GRID CONNECTION & PPAS

Map of DNOs
tinyurl.com/map-dnos

PPA briefing (US-focused but mostly applicable)
tinyurl.com/ppabriefing

SHOWCASE BUILDINGS

BRE Innovation Park
bre.co.uk/innovationpark

Sustainable Energy Academy
sustainable-energyacademy.org.uk